SpringerBriefs in Applied Sciences and Technology

SpringerBriefs present concise summaries of cutting-edge research and practical applications across a wide spectrum of fields. Featuring compact volumes of 50 to 125 pages, the series covers a range of content from professional to academic.

Typical publications can be:

- A timely report of state-of-the art methods
- An introduction to or a manual for the application of mathematical or computer techniques
- A bridge between new research results, as published in journal articles
- A snapshot of a hot or emerging topic
- An in-depth case study
- A presentation of core concepts that students must understand in order to make independent contributions

SpringerBriefs are characterized by fast, global electronic dissemination, standard publishing contracts, standardized manuscript preparation and formatting guidelines, and expedited production schedules.

On the one hand, **SpringerBriefs in Applied Sciences and Technology** are devoted to the publication of fundamentals and applications within the different classical engineering disciplines as well as in interdisciplinary fields that recently emerged between these areas. On the other hand, as the boundary separating fundamental research and applied technology is more and more dissolving, this series is particularly open to trans-disciplinary topics between fundamental science and engineering.

Indexed by EI-Compendex, SCOPUS and Springerlink.

Juan Moreno Nadales · David Muñoz de la Peña ·
Daniel Limon · Teodoro Alamo

Optimal Vessel Planning
in Natural Inland Waterways

 Springer

Juan Moreno Nadales
Ingeniería de Sistemas y Automática
University of Seville
Sevilla, Spain

David Muñoz de la Peña
Ingeniería de Sistemas y Automática
University of Seville
Sevilla, Spain

Daniel Limon
Ingeniería de Sistemas y Automática
University of Seville
Sevilla, Spain

Teodoro Alamo
Ingeniería de Sistemas y Automática
University of Seville
Sevilla, Spain

ISSN 2191-530X   ISSN 2191-5318 (electronic)
SpringerBriefs in Applied Sciences and Technology
ISBN 978-3-031-64743-7   ISBN 978-3-031-64744-4 (eBook)
https://doi.org/10.1007/978-3-031-64744-4

This Springer imprint is published by the registered company Springer Nature Switzerland AG
The registered company address is: Gewerbestrasse 11, 6330 Cham, Switzerland

If disposing of this product, please recycle the paper.

Contents

Chapter 1
Introduction

1.1 Introduction

This book delves into the intricate world of scheduling and rescheduling vessels in natural inland waterways, offering insights and strategies crucial for effective management in this dynamic environment. Inland waterways serve as vital arteries for transportation, facilitating the movement of goods and people across regions. However, managing vessel schedules amidst varying conditions poses significant challenges. From weather fluctuations to unexpected maintenance issues, navigating these waterways demands meticulous planning and adaptability. Within these pages, we explore the complexities of scheduling within the context of natural inland waterways. Drawing from both theoretical frameworks and practical experiences, we dissect the key factors influencing scheduling decisions and delve into innovative approaches for optimization.

1.2 Motivation and Objectives of This Book

The management of maritime transportation and logistic systems has traditionally been a complex, poorly automated task where a significant burden of decision-making has rested on expert operators responsible for performing each of the various tasks. Despite the expertise of these operators in efficiently carrying out each particular piece of work, this management approach comes with a series of problems and inconveniences. On one hand, we encounter a lack of efficiency and optimality when making decisions, which is inherent to human nature when it comes to decision-making. For instance, it is difficult to carry out an optimal overall management of vessels sailing simultaneously in the canal, since vessel planning is carried out on an individual basis. For this reason, it is necessary to provide each of the operators with the necessary decision support tools to carry out their work as efficiently as

J. Moreno Nadales et al., *Optimal Vessel Planning in Natural Inland Waterways*,
SpringerBriefs in Applied Sciences and Technology,
https://doi.org/10.1007/978-3-031-64744-4_1

1

possible while minimizing the probability of error and suffering and accident. On the other hand, the high degree of specialization of each operator in their particular task, along with their limited knowledge of other tasks, greatly complicates the ability of all actors to act in a coordinated manner. For example, an operator responsible for the storage and stacking of cargo containers may not know the most suitable way to organize the containers for optimal stowage. This clearly calls for a change in approach that allows for centralized and coordinated planning of the different tasks involved in the logistics process, with a focus on achieving the overall optimality of the operation.

In addition to these two problems, there are a series of added inconveniences associated with information handling in the maritime community. So far, all data recording in each stage of the process has been done manually and in a rudimentary manner. For instance, the arrival times of each vessel at some specific port or the recording of the type of cargo carried by each vessel has been recorded on paper or communicated via phone between different agents.

In this context, the different measures adopted to address these issues can be grouped into two different lines of action. On one hand, the continuous improvement, modernization, and adaptation to the current requirements and demands of logistics and port infrastructures. On the other hand, the updating of management systems for the different tasks involved in the logistic process through the adoption of measures focused on the use of new technologies. This can result in greater overall process efficiency, a significant reduction in operational costs, improved safety, and a reduced environmental impact related to resource optimization and a significant decrease in energy consumption.

Among all the fields of action where a significant improvement in efficiency can be achieved in the area of maritime freight transport, this book focuses on the research, analysis, development, and implementation of strategies and tools to improve navigation in natural inland waterways, which are natural rivers that have not been artificially modified by human action and are used as a logistic transport channel. The correct management of shipping in this type of logistic channel not only has a direct impact on the efficiency of the port entities connected through it but also has a strong repercussion on the local economy of the region. For this reason, in recent years, the different public and private agents involved in the different stages and tasks of the logistics transport chain, as well as the other socio-economic agents and administrations have shown great interest in carrying out actions on these maritime trade routes focused in optimizing their management and improve their efficiency and safety.

This book is intended to address some of the main problems associated with scheduling vessels in natural inland waterways:

1. **Implementation of tidal windows searching algorithms to ensure safe navigation in natural inland waterways.**
   Unlike artificial navigation channels, specifically designed or adapted for the transport of goods, transportation in natural inland waterways is heavily influenced by the terrain's characteristics and, in the case of channels with access to

the sea, by the tidal dynamics. The irregular bathymetric profile of the channel bottom, coupled with the changing effects of tides, results in one of the primary challenges in conducting transportation in this type of channels: the variation in both space and time of the channel's depth. This compromises navigation safety due to the risk of grounding accident. In general, it is challenging to precisely determine the riverbed's topography at each location, primarily due to the significant influence of sediment transport. Additionally, in most natural inland waterways, especially those that pass through protected areas or landscapes, dredging activities for bottom leveling are often highly restricted. Furthermore, the difficulty of obtaining dynamic models that accurately depict the tidal conditions at any given moment exacerbates the situation. In most cases, the solution is to establish a network of depth sensors at various sections of the river, but the measurements taken at each point are not sufficient to ensure navigation safety in all channel locations. To address this issue, this work aims at proposing algorithms for finding safe crossing windows at different river sections through a reachability analysis, ensuring the existence of safe routes that vessels can follow to sail through the waterway without violating neither speed nor depth constraints.

2. **Implementation of navigation scheduling and planning algorithms in natural inland waterways.**

   When it comes to managing any type of inland waterway, the main challenge is to carry out proper trip planning for vessels within the canal. Taking into account the arrival times of vessels at the estuary, along with other characteristics and requirements related to the arrival time constraints at their destination is a challenging task, as it is necessary to plan the access times of each vessel to the estuary and the speed at which the vessels must navigate through it. To address this planning problem, it is necessary to consider, in addition to the safe navigation time windows imposed by the varying depth effect, all operational constraints placed on each vessel, such as their maximum speed, as well as interactions with other vessels, including crossings and overtaking maneuvers. To address this problem, this work aims to propose a methodology for formulating an optimization problem to optimize the planning of vessels in natural inland waterways taking into account all of these constraints. The goal is to optimize a specific navigation performance index. The adopted criterion can vary from minimizing waiting and transit times in the estuary to minimizing navigation speed to ensure minimal fuel consumption. Along with the methodology for searching safe navigation windows, vessel adherence to the found navigation plans ensures optimal and safe navigation.

3. **Development of dynamic rescheduling methodologies for optimal incident management.**

   Applying and adhering to the navigation plans obtained during the scheduling stage is a challenging task due to the numerous sources of uncertainty and unexpected events that natural inland waterways often encounter. Most of the current planning strategies are often not robust in the face of unexpected events. Additionally, the rescheduling of vessels in these types of channels often results in high administrative costs due to non-compliance with the agreements adopted. Furthermore, causing harm to certain vessels due to incidents beyond their control or related to the poor operation of other vessels can seriously damage the reputation and positioning of the port. Any deviation from the original plan for a vessel can potentially lead to a grounding accident or even a collision with another vessel. For this reason, it is imperative to implement strategies that allow for the rapid detection and assessment of any occurrence during the navigation of each vessel. This work aims to carry out the implementation of a monitoring and accident detection system that allows detecting in real time any possible deviation of the trajectory of the vessels with respect to the original established plan. The proposed method must also be capable of carrying out the correct classification of the type of incident produced in order to adopt the necessary measures for optimal and safe rescheduling. Once the accident has been detected and classified, it is necessary to propose different rescheduling strategies that allow reaching a compromise between the number of vessels affected by the rescheduling and the overall cost of the operation.

4. **Practical implementation of decision support tools for scheduling vessel in natural inland waterways.**

   To carry out the optimal management of an inland waterway, it is important to provide the operators responsible for canal planning with tools that enable them to perform this task in a straightforward manner. Such tools should be able to integrate planning algorithms specifically designed to meet the unique requirements of each channel. In this work, we present the implementation of a software tool designed to address these challenges. The implemented system focuses on optimizing the planning of cargo vessel journeys in inland waterways. This software tool, along with the integrated library of functions, equips navigation planners with a comprehensive solution for optimizing vessel routes while considering the constraints imposed by varying depths and encounter situations. Specifically, the tool must implement two main functions. On the one hand, we need to calculate safe paths to ensure the existence of feasible routes that do not violate any of the imposed restrictions. Secondly, the tool must be able to implement scheduling algorithms for the optimal planning of navigation in the channel. Furthermore, in order to provide the system with greater flexibility, all the information required by the tool to carry out the planning must be stored in a cloud-based database, so that all information can be accessed in real time through a query from the platform where the tool is integrated.

1.3   Natural Inland Waterways

Managing river ports poses a complex challenge due to the intricate nature of coordinating various tasks and resources. The berth allocation problem focuses on optimizing the assignment of berths for incoming ships, ensuring a streamlined flow of vessels. Quay crane scheduling tackles the efficient scheduling of cranes responsible for loading and unloading containers onto and from ships, enhancing overall operational efficiency. Yard crane scheduling addresses the movement and storage of containers within the port's yard, optimizing crane utilization. Tubgoat scheduling addresses the optimal planning of tugboats, crucial for assisting ships in navigating the river. The container stacking problem involves the strategic placement of containers in the yard to maximize space and accessibility. The container drayage problem focuses on planning the transportation of containers between the port and its inland destinations. These challenges underscore the need for sophisticated solutions in port management to minimize delays and enhance overall effectiveness.

Within the broad spectrum of management problems in river ports, the focus of this book is to concentrate on the challenge of optimal vessel planning in this type of channels. The objective is to provide detailed insights and information specifically related to this aspect, addressing the complexities and strategies associated with ensuring efficient and effective navigation within natural waterways connected to inland ports. In order to exemplify the problems associated with navigation in natural inland waterways, this book has taken the case of the Guadalquivir River as an example.

1.4 The Case of the Guadalquivir River

The Guadalquivir River, flowing through the heart of Andalusia in southern Spain, is one of the Iberian Peninsula's major waterways. Originating in the Sierra de Cazorla mountains, the river embarks on a journey of over 650 km, meandering through diverse landscapes before emptying into the Gulf of Cádiz.

Beginning at the Port of Seville, the Guadalquivir River becomes a conduit for maritime activities, connecting the city to the vast network of waterborne trade. The port has been a strategic hub for centuries, facilitating the transport of goods and fostering cultural exchanges. Today, as one of Spain's principal river ports, Seville's harbor accommodates a diverse array of vessels, reflecting the river's enduring role in regional commerce. A satellite image of the Guadalquivir River from the Port of Seville to the river's estuary is depicted in Fig. 1.1. As in most natural inland waterways, navigation in the Guadalquivir River is significantly influenced by two key factors: the effect of tides and the bathymetric profile of the watercourse. The influence of tides, with their constant ebb and flow, creates a dynamic and variable environment that directly impacts maritime navigation. The height of tides and their periodic changes affect the river's depth, posing challenges for navigation, especially in narrower stretches and shallow areas. A detailed description of the effect of the tide on the depth of the river can be found in [1]. Examples of tide measurements at different locations along the Guadalquivir River are shown in Fig. 1.2. The sensor locations are not displayed for confidentiality reasons.

Furthermore, the bathymetric profile, or the configuration of the riverbed, adds another layer of complexity. This problem is particularly acute in canals that flow through areas of high environmental protection, as is the case of the Guadalquivir River in the Doñana National Park, since dredging operations are highly restricted. A detailed description of the bathymetric profile in the Guadalquivir River can be

Fig. 1.1 Satellite image of the Guadalquivir River connecting the city of Seville and the Atlantic Ocean as it passes through Doñana National Park

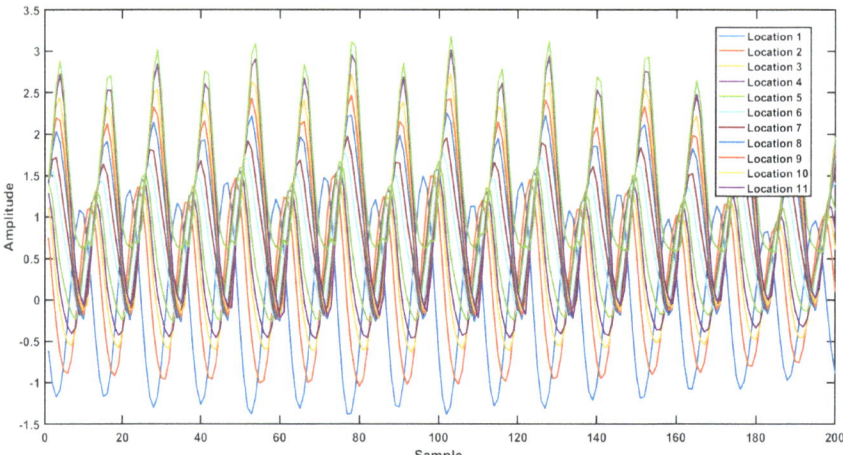

Fig. 1.2 Amplitude of the tide wave at 11 different locations of the Guadalquivir River

found in [2]. Changes in the depth and shape of the riverbed determine the tidal hydrodynamic regime along the river. Variations in the width and depth of the water body affect the behavior of tidal currents and the distribution of tidal energy. This implies that certain sections of the river may experience stronger currents, while others may have calmer waters.

Collectively, the interaction between the tidal effect and the bathymetric profile creates a complex scenario for navigation on the Guadalquivir. Navigators must take into account tidal fluctuations and adapt to the changing conditions of the riverbed to ensure safe and efficient navigation. A detailed understanding of these factors is essential for developing effective river management strategies and ensuring the sustainability of navigation activities on this fascinating watercourse.

In addition to the aforementioned problems associated with various natural phenomena such as tides and irregular bathymetric profiles, managing navigation in

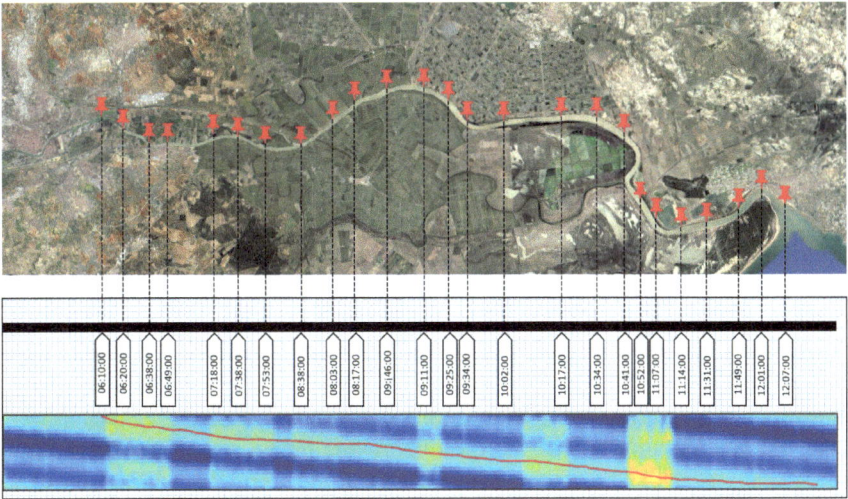

Fig. 1.3 Example of trip plan. The depth map is shown at the bottom. The darker areas represent the areas where the depth is shallower

natural inland waterways requires consideration of operational constraints related to ensuring navigation safety and potential interactions among different vessels. Thus, it is necessary, for example, to account for the maximum speeds at which vessels can travel through each section of the canal, the minimum distance between two vessels navigating in the same direction, or the optimal crossing location between two vessels traveling in opposite directions, considering their dimensions and the width of the canal.

At present, the management process of navigation in most inland waterways, as in the case of the Guadalquivir River, is solved in a rudimentary way. Experts in the management of the channel carry out a daily planning of the trips of the different vessels. This is translated into a navigation plan indicating the moments of time in which each of the vessels must pass through each of the different sections of the river. An example of trip plan is depicted in Fig. 1.3. This figure shows each of the time instants in which a given vessel has to cross the different boundaries that divide the river into segments. From here, the pilot in charge of the control of the boat elaborates a navigation plan to fulfill the planning without violating any of the imposed restrictions, giving rise to the following of a trajectory as shown in red in the lower part of the figure, where it can be observed how the restriction imposed by the depth of the river is not violated at any moment. Both in the design of the planning of the boat trips on the river, as well as in the management of the same by the pilots, there are many risks due to the human component in the decision-making process, putting at risk the optimality and safety of the operation. This problem requires the development of optimal planning strategies to ensure the safety of navigation and mechanisms to deal with the occurrence of incidents associated with the human component of the process.

1.5 Background

1.5.1 Scheduling in Natural Inland Waterways

In the consumer society we live in, maritime transport is one of the sectors with a greater impact on countries' development and the growth of global economy [3]. As it has been demonstrated [4], the pace of growth of a country's economy is directly and highly correlated with the amount of investment on logistic infrastructures and the improvement of maritime transport. Thus, it is of utmost importance to design a road map to adapt existing port systems to meet the requirements of today's society. This requires not only the modernization of port facilities and the use of new technologies [5], but also the development of more efficient management strategies to improve the efficiency of the overall logistic process [6].

Many research lines have focused on improving the efficiency of different port operations. This is the case of the berth allocation problem (BAP) [7–10], which deals with optimizing berthing times, or the quay crane scheduling problem (QCSP), which aims at scheduling quay cranes for loading and unloading containers in such a way that vessels' turnaround times are minimized [11–15].

In addition, in inland ports connected to the sea coast through some channel or waterway, as in the case of the Port of Seville, the navigation-oriented trip planning of vessels plays a key role in port efficiency. This problem is known as the waterway ship scheduling problem, whose objectives include deciding the optimum entry times into the navigation channel for each of the vessels, as well as the guidelines to be followed by the vessel pilots to navigate optimally through the channel. This problem does not only need to consider matters relating to vessels' characteristics, but also different exogenous factors that may affect the waterway, such as the effect of the tide on the depth. Due to its great importance in the current economics of maritime transport in this type of channels, several studies have been carried out in the academy during the last years dealing with this problem.

In [16], authors aim at enhancing the efficiency of vessel transportation through vessel scheduling in inland ports by effectively coordinating the use of channels and berths. It establishes a mathematical model with the objective of minimizing the total waiting time, considering factors such as scheduling order, travel direction, and berth distance. To address this scheduling challenge, the study employs simulated annealing and multiple population genetic algorithm. By comparing the results to conventional methods, the study demonstrates that the proposed approach significantly reduces waiting times, scheduled time, and maximum waiting time, simultaneously ensuring both safety for vessels and efficiency.

In [17] authors introduce a ship traffic scheduling model designed to significantly improve traffic efficiency in constrained waterways. The model assumes that two-way traffic is not allowed when large ships are present. It proposes a mathematical model aimed at minimizing the weighted average of mean and maximum waiting times while considering safety restrictions. To solve this problem, a sequential scheduling algorithm is presented, where ships are organized into rounds, and ships within each round are scheduled simultaneously. This approach enhances traffic efficiency by balancing the priorities between small and large ships. The article also includes experiments comparing the sequential scheduling method with a first-come-first-serve model. The results show that the proposed algorithm reduces waiting times for ships, particularly in cases with a higher proportion of large vessels. It ensures that ships' waiting times remain at an acceptable level, whereas the first-come-first-serve approach results in long waiting times.

In [18] authors develop a fuzzy scheduling optimization method for one-way waterways, aiming to enhance port competitiveness by minimizing ship waiting times, improving traffic efficiency, and accommodating uncertain factors. It extends previous research on one-way waterways scheduling by introducing triangular fuzzy numbers to address ship speed and crossing time times, especially considering tide height constraints. The research employs a genetic algorithm to construct a mathematical model for fuzzy vessel scheduling based on time optimization, and it uses the minimum delay strategy for service sequence determination. The study conducts experimental comparative analyses and concludes that this fuzzy vessel scheduling algorithm enhances efficiency and traffic safety, reduces dependence on port conditions, and avoids issues encountered with manual scheduling methods.

Other research lines have focused on solving all the different problems affecting the traffic scheduling problem for some particular waterway or port to provide a global solution, such as the Port of Shanghai. For this, the authors of [19] address the challenge of efficiently managing vessel traffic in a seaport to minimize delays and the number of vessels unable to berth or depart due to congestion. It proposes an approach that utilizes a mixed-integer linear program (MILP) and a set-partitioning formulation to optimize traffic scheduling, along with a column generation algorithm and performance enhancement strategies. Computational results were evaluated using data from a specific port, highlighting the effectiveness of the proposed algorithm in reducing computation times. Furthermore, future extensions of the model are suggested, including the integration of berth allocation decisions and the consideration of uncertainties in port operations. As an extension of this work, the authors of [20] address the optimization of capacity and availability in navigation channels and anchorage areas within a container port. The focus of the article is to develop a mathematical model that simultaneously optimizes traffic in the navigation channels and the utilization of anchorage areas. A mixed-integer programming formulation of the problem is presented, its complexity is analyzed, and a Lagrangian relaxation heuristic is proposed. The computational performance of this heuristic is evaluated on problem instances generated from operational data of a port in Shanghai. The obtained results indicate that the proposed heuristic achieves satisfactory performance within a reasonable computation time, further advancing the research in the efficient management of maritime traffic in ports. In [21], a step further is taken, and navigation management is done taking pilot assignments into account. This article builds upon previous research and addresses the integrated problem of scheduling vessel traffic and pilot assignments in the busiest seaports. It focuses on optimizing the utilization of navigation channels and anchorage areas in the port terminal, aiming to mitigate congestion and enhance vessel service. The problem is formulated on a time-space network with costs dependent on both vessels and pilots, and an integer programming model is developed to minimize various associated costs. To solve the problem, feasible vessel paths are enumerated in advance, and the Lagrangian relaxation algorithm proposed in the previous work is employed to decompose the problem into subproblems for vessel and pilot path assignments. The computational performance of the Lagrangian relaxation algorithm is evaluated using problem instances generated from the physical layout and operational data of the Waigaoqiao Port in Shanghai. This approach, which integrates the vessel traffic management problem with the pilot planning problem, is also treated in [22]. The latter focuses on the pilotage planning problem in seaports, which involves scheduling vessel traffic, assigning work shifts to pilots, and scheduling pilots for vessel navigation. The problem is formulated as a mixed-integer linear programming model with the objective of minimizing the cost associated with sailing operations. The study shows that this problem is strongly NP-hard. To solve it, a branch-and-bound algorithm is developed, featuring a novel dynamic programming algorithm for pricing. Several acceleration techniques are proposed to enhance the algorithm's efficiency. Computational experiments reveal that the branch-and-bound algorithm is capable of solving practical-sized problem instances and outperforms both standard commercial solvers .

This work is complemented by the work developed in [23], where a problem for scheduling tugboats for mooring and unmooring operations in busy seaports is proposed. In a busy port with a limited number of tugboats, effective scheduling of tugboats is crucial for the execution of vessels' berth plans. This paper addresses the tugboat scheduling problem, utilizing a network representation and an integer programming formulation that considers vessel berth plans, tugboats, and power requirements. The objective is to minimize the weighted sum of vessel berthing and departure lateness, tugboat operating costs, and the number of vessels that cannot be successfully served. The authors also introduce a novel iterative solution method involving Lagrangian relaxation and Benders decomposition for generating near-optimal solutions. Computational experiments assess the performance of this solution method using operational data from a container port in Shanghai.

In the same vein, in [24], authors propose a scheduling problem that aims at addressing inter-shipping line equity in the navigation channel of a seaport while trying to find the best compromise between two conflicting objectives, equity and efficiency, again considering tidal windows to access and sail through the channel. Unlike traditional approaches focused solely on minimizing vessel delays, this study prioritizes addressing inter-shipping line fairness. An innovative lexicographic optimization model is developed, balancing the conflicting goals of efficiency, favoring total vessel delay reduction, and equity, ensuring a fair distribution of delays among shipping lines. The model allows port operators to quantify the trade-off between efficiency and equity and make informed scheduling decisions. To solve this problem, a two-stage solution method is presented. First efficiency and equity are optimized separately. Then the best trade-off between both objectives is determined. The model and solution method are applied to real-world data from the Port of Shanghai, and computational results reveal that an equity-oriented approach can achieve satisfactory service at both the vessel and shipping line levels with minimal losses in efficiency, demonstrating the significance of considering inter-shipping line equity in vessel traffic scheduling.

Similar cases can be found in other locations, each with the unique characteristics and circumstances of each waterway. An example could be the numerous works that can be found in the literature focused on the Kiel Canal, the world's busiest artificial waterway. Under this scenario, in [25], authors introduce a practical optimization problem known as the ship traffic control problem. The canal operates bidirectionally, with large ships only able to pass each other in designated sidings. The objective is to minimize the total waiting times of all ships, necessitating complex decisions about ship priorities, waiting locations, and duration, considering various operational constraints. The authors integrate algorithmic concepts from collision-free routing of automated guided vehicles, providing a unified approach to scheduling and dynamic routing. Their traffic control tool, utilizing combinatorial algorithms, significantly improves scheduling over manual planning and plays a vital role in a canal enlargement project to ensure continued operability as traffic volume and vessel sizes increase. In [26], the authors address the ship traffic management challenge in the same canal, focusing on efficient and safe transits. The paper introduces a set of optimization models that consider factors such as variable ship speeds, siding segment capacities, and ship waiting time limits, while adhering to traffic rules and safety requirements. The primary objective across these models is to minimize the total transit time for ships. The authors propose a metaheuristic approach and validate its excellent performance through experiments with real-world data, underscoring its potential to provide high-quality service for ships using the canal.

However, regardless of the great interest that this problem has raised, most works focusing on natural waterways do not pay particular attention to the effect of natural phenomena such as the irregular bathymetric profile and tidal effect, which complicates the problem including dynamic constraints. Regarding this issue, in [27], the authors address the waterway ship scheduling problem in the Yangtze River, which is affected by tidal effects. The proposed problem aims to schedule ship traffic along waterways to minimize waiting times, thus addressing congestion and reducing emissions. The authors propose a mixed-integer linear programming model implemented using CPLEX, along with the introduction of greedy heuristics and a simulated annealing (SA) algorithm. Computational experiments reveal the inherent complexity of the problem, even with the utilization of a general-purpose solver, while demonstrating that the simulated annealing approaches provide efficient solutions in a short computational time. Comparisons with common queue policies highlight the substantial improvements in reducing unnecessary ship waiting times, resulting in enhanced traffic flow, economic benefits, operational efficiency, and environmental savings. This study emphasizes the adaptability and practical relevance of SA, especially for scenarios affected by tidal effects, offering advantages for terminal competitiveness, cost reduction, and expedited negotiations between terminal operators and shipping companies by quickly generating feasible schedules without compromising performance. This problem is revisited by the same authors in [28]. There, the authors propose a reformulation of waterway ship scheduling problem as a variant of the multi-mode resource-constrained project scheduling problem, incorporating time-dependent resource capacities and time constraints. They solve this problem using integer programming, achieving optimality for all instances in the existing literature. The paper offers a dual contribution by introducing the multi-mode resource-constrained modeling approach, transferable to maritime logistics applications, and by enhancing results for real-world-based instances. Computational studies demonstrate the benefits of their approach over previous methods, suggesting its suitability for real-time decision support environments.

In [29] the authors tackle the issue of efficient vessel scheduling through a one-way tidal channel, with a focus on improving port operational efficiency and addressing delays caused by tidal cycles. They present a model and algorithm designed to optimize vessel scheduling, ensuring the shortest total waiting time for vessels entering the channel. The model and algorithm are validated using real data from the Yangtze River estuary, demonstrating their superior scheduling efficiency compared to alternative strategies, including first-come, first-served, random scheduling, larger draft vessels priority scheduling, and manual scheduling. Furthermore, the adaptable nature of the proposed model and algorithm allows for their application to vessel scheduling in other channels with tide-affected depth. Extensions of this work for two-way and compound tidal channels are proposed in [30, 31], respectively.

In [32], the authors address the challenges of scheduling vessels on restricted waterways due to the increasing size of vessels and the limited possibilities for waterway expansion. The study focuses on a tide-dependent waterway scheduling problem with a passing box, where wide vessels can only pass within the passing box, and vessels with deep drafts are constrained to specific time windows around high tide. The authors present a mixed-integer program for this problem and introduce techniques to enhance the model formulation by minimizing Big-M values, fixing variables, and reducing the number of variables and constraints. The proposed model formulations are evaluated through a comprehensive computational study using real-world data from the Elbe River near Hamburg, Germany. The results indicate that the passing box's capacity is not restrictive in the studied setting. However, further research is needed to explore the implications of smaller passing boxes, increased passing box capacity, and the development of effective heuristics and meta-heuristics for solving tide-dependent waterway scheduling problems, especially when multiple passing boxes are considered. The refined model formulation can serve as a valuable resource for researchers designing heuristic solutions for such problems.

Finally, and given its significant relevance to the present work due to its application to the Guadalquivir River, the reader is referred to [33], where the planning problem in the Guadalquivir waterway is addressed. This paper addresses the complex task of planning vessel navigation along a tidal river, which significantly impacts the efficiency of inland ports and intermodal chains. The study introduces a heuristic procedure designed to schedule vessels on a tidal waterway, aiming to optimize navigation based on a multi-criteria objective function. This function combines various factors, including the number of vessels serviced, total waiting time, waterway occupation time, and the count of crossing operations. The procedure incorporates depth estimation to identify critical navigation points and potential bottlenecks, considering scenarios where vessels may need to anchor and wait for the next tidal window. The proposed strategy is validated through application to simulation scenarios, demonstrating reliable results in terms of performance and solution quality.

1.5.2 Safety-Oriented Rescheduling in Natural Inland Waterways

While a substantial body of research has focused on the planning of vessels in inland waterways, the majority of these studies primarily center on developing scheduling plans optimized for ideal operational conditions. As a result, they may not guarantee optimal outcomes and the safety of various operations when unforeseen events, such as delays or mechanical breakdowns, occur. In recent years, some scholars have advocated for considering decision-making uncertainty, leading to complex optimization challenges, as reflected on the following examples.

The importance of proper management of navigation safety in inland waterways is becoming evident and is reflected in the high level of interest generated within academia for developing safety management strategies and accident risk assessment in this type of channels. In [34], the authors introduce an innovative and multi-faceted safety evaluation model for the navigation of ships in inland waterways. This model operates at a three-level framework, incorporating advanced principles of fuzzy theory and evidential reasoning. The three-level structure integrates various safety indicators and takes into account both qualitative and quantitative data, resulting in a robust assessment of navigation safety. This pioneering model represents a significant milestone in the field of maritime safety management. It plays a crucial role in preventing potential accidents and providing early warnings for inland waterway transportation. One of the standout features of this research is its recognition of the immense value of real-time monitoring and the utilization of real-time data. Despite occasional gaps in data availability, the research underscores the pivotal role played by real-time information in enhancing safety management. Moreover, it highlights the model's potential to seamlessly integrate with waterway traffic management platforms. The essential data needed for this model's operation is sourced from the automatic identification system (AIS) [35]. AIS is a transponder technology that automatically provides vital information, including vessel position data, vessel identification, and other pertinent details, to both fellow vessels and coastal authorities. This comprehensive and multi-level safety evaluation model has the potential to enhance safety practices in inland waterway navigation, making it a valuable asset in the realm of maritime safety management and navigation risk mitigation.

Another noteworthy study emphasizing the significance of real-time monitoring and the utilization of AIS data for enhancing navigation safety management in inland waterways can be found in [36]. This article delves into the technical intricacies of enhancing navigation for inland waterway vessels through driver-assistance functionalities, with a primary focus on the architectural framework for the delivery of precise and dependable position, navigation, and timing data. The proposed approach integrates AIS communication protocols to facilitate the transmission of global navigation satellite system (GNSS) correction data. The navigation algorithm employs code and phase GNSS observations in conjunction with differential correction data to accurately estimate positioning with centimeter-level precision. Furthermore, an integrity monitoring algorithm is implemented to ensure data reliability. The research underscores the critical importance of GNSS correction data transmission and explores the potential utilization of AIS as a communication medium.

The importance of addressing unexpected events in maritime operations cannot be overstated. These events can significantly impact the optimality and safety of vessel operations, which are critical in natural estuaries. When we look at the planning algorithms that have been developed in this domain, it becomes evident that most of them lack robustness when faced with the inherent uncertainty associated with the diverse range of scenarios that can unfold during a vessel's journey through a natural estuary. These scenarios can include adverse weather conditions, sudden equipment failures, changes in traffic congestion, and unexpected navigational hazards, among others. These unforeseen events demand a more comprehensive and dynamic approach to planning and safety management, one that can adapt in real time to minimize the risk of accidents and maintain operational efficiency. Many existing replanning strategies predominantly focus on re-routing vessels between logistic nodes. While this is an essential aspect of maritime operations, it does not adequately address the dynamic challenges vessels encounter along their entire route through a canal. These challenges may involve rapidly changing conditions that necessitate continuous monitoring and decision-making.

To overcome these limitations, it is imperative to develop replanning methodologies and safety management systems that are capable of addressing incidents as they occur in real time. This requires a holistic approach to maritime operations that integrates the principles of safety management, risk assessment, and dynamic replanning. One key aspect of these new methodologies is the incorporation of advanced sensor technologies and data analytics. These technologies can provide real-time data on vessel position, weather conditions, water depth, and other critical parameters. By analyzing these data, algorithms can continuously assess the operational context and potential risks, allowing for adaptive decision-making. Moreover, the human element in maritime operations should not be overlooked. Crew members play a crucial role in ensuring the safety and efficiency of operations. Hence, training and education programs should be designed to equip them with the skills to handle unexpected events effectively and make informed decisions under pressure. In addition to the technical and human factors, a regulatory framework must be established to ensure that these rescheduling methodologies and safety management systems are implemented effectively across the maritime industry. Regulations should set standards for real-time incident response, vessel monitoring, and reporting procedures.

In [37] the authors delve into the intricacies of ship scheduling within the Kiel Canal, specifically addressing the challenges posed by the uncertainty of ship arrival times and the resulting frequent need for schedule adjustments to ensure feasibility. To mitigate these issues, the authors propose a sophisticated mathematical model that incorporates the concept of time-corridors. These time-corridors act as flexible windows within which ships must arrive, allowing for dynamic scheduling while maintaining the validity of the plans. To efficiently solve this complex problem in practical scenarios, they introduce a heuristic approach that systematically minimizes the necessity for rescheduling. This approach is compared to the prevailing simple waiting heuristic, revealing its superiority in reducing both the frequency of daily reschedules and the overall time it takes for ships to traverse the canal. Furthermore, the study meticulously explores the determination of optimal time-corridor widths, offering insights into the trade-off between minimizing reschedules and ensuring efficient ship transit times. It also investigates the influence of notice time, which represents the advance notice ships provide regarding their estimated arrival times, on the scheduling process. The results show that notice time has a significant impact on rescheduling frequency, underscoring the importance of this parameter in canal scheduling operations. Additionally, the article scrutinizes the effects of upgrading narrow transit segments along the canal. The findings indicate that enhancing these segments, particularly when done collectively, can yield substantial reductions in the average traversing times for ships. In summary, this comprehensive research highlights the advantages of using time-corridors to address scheduling uncertainties, reduce rescheduling requirements, and improve overall efficiency in managing ship traffic through the Kiel Canal.

In [38], the authors address the challenging problem of ship sequencing and scheduling in the context of Yangtze River traffic management, especially in restricted waterways. The authors propose an innovative sliding window-based online ship sequencing and scheduling algorithm (OSS-SW) to tackle this issue effectively. The OSS algorithm introduces the concept of position shift, capitalizing on differences in ships' sailing times within the restricted waterway to generate a more efficient ship sequence. The sliding window mechanism is incorporated to handle traffic uncertainties and reduce computational complexity. The study also investigates the impact of congestion in restricted waterways on the performance of the OSS-SW algorithm and thoroughly explores parameter settings. Through both simulation studies and real data applications, the authors demonstrate that the proposed OSS-SW algorithm outperforms existing methods, including the traffic signal revealing system (TSRS) and the expert system-based algorithm (ES), in solving ship sequencing and scheduling challenges in the Yangtze River traffic management. The paper proposes a novel approach by incorporating the sliding window scheme to reduce computational burden and improve solution quality under various uncertain scenarios. The experiments confirm that the proposed OSS-SW method achieves similar or better performance than the OSS-only method and outperforms TSRS and ES methods. The algorithm demonstrates its effectiveness in obtaining optimal ship sequences for both upstream and downstream traffic, significantly reducing computational complexity. Additionally, Monte Carlo simulations reveal the algorithm's robustness under different scenarios and parameter settings, suggesting that it can handle ship sequencing and scheduling challenges effectively in restricted waterways. The study concludes by highlighting the potential for further development of online optimization methods in the future to enhance inland waterway capacity, such as the Yangtze River, despite the inherent complexity of the ship sequencing and scheduling problem.

In [39], the authors focus on the problem of optimizing vessel speeds, particularly when faced with uncertainties related to waiting times at locks due to unpredictable processing time estimations of other vessels. The primary objective is to minimize fuel consumption for approaching ships while adhering to predefined deadlines for traversing river segments. To tackle this challenge, the authors introduce a sophisticated mathematical optimization model that takes into account the uncertainties inherent in lock operations. This mathematical model forms the basis for two distinct solution approaches: an optimal solution employing complex mathematical algorithms to determine the most fuel-efficient speeds, and a simplified heuristic offering practical guidance for skippers to make informed speed choices that balance fuel efficiency with transit times. These approaches are developed to empower skippers with effective strategies for decision-making amid unpredictable lock waiting times, ultimately aiming to enhance operational efficiency and sustainability in inland waterway transportation.

1.6 Organization of This Book

This book is organized as follows:

- In Chap. 2, the entire methodology for the planning of trips of vessels in natural inland waterways is developed. First of all, we introduce the problem of ship planning in natural inland waterways, introducing the problem to be solved. Next, the algorithm for calculating safe navigation tubes is presented, detailing each and every one of the different processing stages for obtaining the safe crossing windows from the depth map. This part presents the tree algorithm based on reachability analysis for concatenating safe crossing windows at each boundary of the waterway and the pruning algorithm for the determination of the maximum and minimum crossing times for each boundary according to the constraints imposed by the depth map and the maximum velocity at each location of the depth map.
Subsequently, once the algorithm for the calculation of safe tubes for each vessel has been developed, an optimization problem formulation is presented to obtain the optimal crossing times when each vessel should cross each boundary in the waterway. In addition to the constraints imposed by the tide, the rest of the operative constraints are presented, as well as the performance index to minimize the overall waiting and navigation times. Finally, the safe tube calculation algorithm and the optimization problem presented are used in simulation in realistic case studies using real data from the Guadalquivir estuary, comparing how the proposed methodology improves the current first-arrived/first-served managing approach.
- In Chap. 3, the monitoring architecture and dynamic rescheduling strategies for incident prevention in natural inland waterways are presented. First, it is proposed a scheme for monitoring navigation in inland waterways, centered on detecting

potential incidents that may jeopardize safety in these types of canals. This scheme is based on comparing the current position with the expected trajectory according to adopted navigation criteria. For the main types of incidents detected in inland waterways, especially in the case study conducted on the Guadalquivir River, we introduce a practical and easily implementable filter for accurate incident type identification. This is crucial for ensuring the proper rescheduling of operations. In order to reschedule vessels when an incident is detected, we compare different strategies for rescheduling the optimal trip plans considering not only efficiency but also the impact and disruption caused by the rescheduling on current plans.

This leads us to adopt a parameterized multi-objective strategy to weight the repercussions on the current planning. For the various rescheduling strategies that are proposed, our focus is on appropriately modifying the planning problem to account for the requirements imposed by the identified incident type. This ensures the proper response to each specific scenario. Different simulations using real case studies in the Guadalquivir River are carried out to demonstrate the effectiveness of the proposed methodology, showing its success in preventing accidents and enhancing optimality.

- Finally, in Chap. 4 implementation of an open-source software tool for the scheduling of vessels in natural inland waterways is presented, taking the Guadalquivir River as an example. This tool implements two main functionalities. On the one hand, it performs the calculation of the safe navigation tubes for each of the vessels taking into account the restrictions imposed by the channel depth and the maximum speed in each location of the river as presented in Chap. 2. Secondly, the optimal trip planning scheduling algorithm is implemented. To this end, a software architecture using Python programming language and a cloud-based data storage system is presented. First, the overall system architecture and the different functionalities of the tool are detailed. Secondly, each and every module and function of the tool is analyzed in detail, showing special interest in the interconnection between the different modules and the management of input and output data of each module. The performance of the tool is shown through a real example of application in the Guadalquivir River. This tool has been made available to the community through a free distribution system. The access procedure is detailed at the end of the chapter.

References

1. S. Sirviente, J. Sánchez-Rodríguez, J.J. Gomiz-Pascual, M. Bolado-Penagos, A. Sierra, T. Ortega, O. Álvarez, J. Forja, M. Bruno, A numerical simulation study of the hydrodynamic effects caused by morphological changes in the Guadalquivir River Estuary. Sci. Total Environ. **902**, 166084 (2023)
2. S. Costa, J. Gutiérrez Mas, J. Morales, Establecimiento del régimen de flujo en el estuario del Guadalquivir, mediante el análisis de formas de fondo con sonda multihaz. Revista de la Sociedad Geológica de España **22**(1–2), 23–42 (2009)

3. N. Akbulaev, G. Bayramli, Maritime transport and economic growth: interconnection and influence (an example of the countries in the Caspian sea coast; Azerbaijan, Turkmenistan, Kazakhstan and Iran). Mar. Policy **118**, 104005 (2020)

4. S.C. Nita, A. Hrebenciuc et al., The importance of maritime transport for economic growth in the European Union: a panel data analysis. Sustainability **13**(14), 7961 (2021)

5. L. Heilig, E. Lalla-Ruiz, S. Voß, Digital transformation in maritime ports: analysis and a game theoretic framework. Netnomics: Econ. Res. Electron. Netw. **18**(2), 227–254 (2017)

6. Y. Jiang, J. Lu, Y. Cai, Q. Zeng, Analysis of the impacts of different modes of governance on inland waterway transport development on the Pearl river: the Yangtze river mode vs. the Pearl river mode. J. Transp. Geogr. **71**, 235–252 (2018)

7. L. Guo, J. Wang, J. Zheng, Berth allocation problem with uncertain vessel handling times considering weather conditions. Comput. & Ind. Eng. **158**, 107417 (2021)

8. K. Buhrkal, S. Zuglian, S. Ropke, J. Larsen, R. Lusby, Models for the discrete berth allocation problem: a computational comparison. Transp. Res. Part E: Logist. Transp. Rev. **47**(4), 461–473 (2011)

9. M.A. Dulebenets, Application of evolutionary computation for berth scheduling at marine container terminals: parameter tuning versus parameter control. IEEE Trans. Intell. Transp. Syst. **19**(1), 25–37 (2018). https://doi.org/10.1109/TITS.2017.2688132

10. X. Lyu, R.R. Negenborn, X. Shi, F. Schulte, A collaborative berth planning approach for disruption recovery. IEEE Open J. Intell. Transp. Syst. (2022)

11. S. Ma, H. Li, N. Zhu, C. Fu, Stochastic programming approach for unidirectional quay crane scheduling problem with uncertainty. J. Sched. **24**(2), 137–174 (2021)

12. D. Sun, L. Tang, R. Baldacci, A. Lim, An exact algorithm for the unidirectional quay crane scheduling problem with vessel stability. Eur. J. Oper. Res. **291**(1), 271–283 (2021)

13. Y. Li, F. Chu, F. Zheng, M. Liu, A bi-objective optimization for integrated berth allocation and quay crane assignment with preventive maintenance activities. IEEE Trans. Intell. Transp. Syst. 1–18 (2020). https://doi.org/10.1109/TITS.2020.3023701

14. R.T. Cahyono, S.P. Kenaka, B. Jayawardhana, Simultaneous allocation and scheduling of quay cranes, Cranes, and trucks in integrated container terminal operations. IEEE Trans. Intell. Transp. Syst. 1–15 (2021). https://doi.org/10.1109/TITS.2021.3083598

15. Y. Zhang, B. Atasoy, R.R. Negenborn, Preference-based multi-objective optimization for synchromodal transport using adaptive large neighborhood search. Transp. Res. Rec. 03611981211049148 (2021)

16. X. Zhang, J. Lin, Z. Guo, T. Liu, Vessel transportation scheduling optimization based on channel-berth coordination. Ocean Eng. **112**, 145–152 (2016)

17. J. Zhang, T.A. Santos, C. Guedes Soares, X. Yan, Sequential ship traffic scheduling model for restricted two-way waterway transportation. Proc. Inst. Mech. Eng. Part M: J. Eng. Marit. Environ. **231**(1), 86–97 (2017)

18. D. Liu, G. Shi, Z. Kang, Fuzzy scheduling problem of vessels in one-way waterway. J. Mar. Sci. Eng. **9**(10), 1064 (2021)

19. S. Li, S. Jia, The seaport traffic scheduling problem: formulations and a column-row generation algorithm. Transp. Res. Part B: Methodol. **128**, 158–184 (2019)

20. S. Jia, C.L. Li, Z. Xu, Managing navigation channel traffic and anchorage area utilization of a container port. Transp. Sci. **53**(3), 728–745 (2019)

21. S. Jia, L. Wu, Q. Meng, Joint scheduling of vessel traffic and pilots in seaport waters. Transp. Sci. **54**(6), 1495–1515 (2020)

22. L. Wu, S. Jia, S. Wang, Pilotage planning in seaports. Eur. J. Oper. Res. **287**(1), 90–105 (2020)

23. S. Jia, S. Li, X. Lin, X. Chen, Scheduling tugboats in a seaport. Transp. Sci. **55**(6), 1370–1391 (2021)

24. S. Jia, Q. Meng, H. Kuang, Equitable vessel traffic scheduling in a seaport. Transp. Sci. **56**(1), 162–181 (2022)

25. E. Lübbecke, M.E. Lübbecke, R.H. Möhring, Ship traffic optimization for the Kiel Canal. Oper. Res. **67**(3), 791–812 (2019)

26. F. Meisel, K. Fagerholt, Scheduling two-way ship traffic for the Kiel Canal: model, extensions and a matheuristic. Comput. & Oper. Res. **106**, 119–132 (2019)
27. E. Lalla-Ruiz, X. Shi, S. Voß, The waterway ship scheduling problem. Transp. Res. Part D: Transp. Environ. **60**, 191–209 (2018)
28. A. Hill, E. Lalla-Ruiz, S. Voß, M. Goycoolea, A multi-mode resource-constrained project scheduling reformulation for the waterway ship scheduling problem. J. Sched. **22**(2), 173–182 (2019)
29. B. Zhang, Z. Zheng, Model and algorithm for vessel scheduling through a one-way tidal channel. J. Waterway Port Coastal Ocean Eng. **146**(1), 04019032 (2020)
30. B. Zhang, Z. Zheng, D. Wang, A model and algorithm for vessel scheduling through a two-way tidal channel. Marit. Policy & Manag. **47**(2), 188–202 (2020)
31. B. Zhang, Z. Zheng, Model and algorithm for vessel scheduling optimisation through the compound channel with the consideration of tide height. Int. J. Shipp. Transp. Logist. **13**(3–4), 445–461 (2021)
32. L. Nehrke, A. Schulz, Scheduling of waterways with tide and passing box. Naval Res. Logist. (NRL) **69**(4), 609–621 (2022)
33. J. Muñuzuri, E. Barbadilla, A. Escudero-Santana, L. Onieva, Planning navigation in inland waterways with tidal depth restrictions. J. Navig. **71**(3), 547–564 (2018)
34. J. Liu, X. Jiang, W. Huang, Y. He, Z. Yang, A novel approach for navigational safety evaluation of inland waterway ships under uncertain environment. Transp. Safety Environ. **4**(1), tdab029 (2022)
35. I.M. Organization, Revised guidelines for the onboard operational use of shipborne automatic identification systems (AIS) (2015)
36. A. Hesselbarth, D. Medina, R. Ziebold, M. Sandler, M. Hoppe, M. Uhlemann, Enabling assistance functions for the safe navigation of inland waterways. IEEE Intell. Transp. Syst. Mag. **12**(3), 123–135 (2020)
37. T. Andersen, J.H. Hove, K. Fagerholt, F. Meisel, Scheduling ships with uncertain arrival times through the Kiel Canal. Marit. Transp. Res. **2**, 100008 (2021)
38. S. Gan, Y. Wang, K. Li, S. Liang, Efficient online one-way traffic scheduling for restricted waterways. Ocean Eng. **237**, 109515 (2021)
39. M. Buchem, J.A.P. Golak, A. Grigoriev, Vessel velocity decisions in inland waterway transportation under uncertainty. Eur. J. Oper. Res. **296**(2), 669–678 (2022)

Chapter 2
Scheduling Vessels in Natural Inland Waterways

Abstract Although crucial to global logistics, inland ports grapple with persistent challenges. Foremost among these is the intricate task of planning cargo vessel journeys. Effective planning not only boosts port efficiency and reputation but is also paramount for operational safety. This challenge is particularly pronounced in natural waterways, where depth and width are subject to uncontrollable natural forces. Take, for instance, the Guadalquivir River, linking the Atlantic Ocean to the inland Port of Seville in the south of Spain. To address this, in this chapter we present a two-step strategy aimed at optimizing vessel journey times amidst varying water depths and encounter scenarios. The proposed solution delineates specific crossing times at boundary points along the waterway. We evaluate its advantages against a traditional first-come, first-served scheduling approach, focusing on optimality and feasibility.

2.1 Introduction

Many approaches to addressing scheduling challenges in artificial or natural watercourses affected by natural occurrences typically overlook the effect of the fluctuating depth along the watercourse. Instead, they primarily concentrate on identifying optimal tidal windows for entering or exiting the channel. In this chapter, we introduce a broader framework for tackling the planning and scheduling dilemmas encountered in inland watercourses. This framework ensures the feasibility of trajectories while considering the diverse array of natural and operational constraints that influence navigation in such watercourses. We propose a two-phase solution approach. Firstly, we identify a sequence of time intervals (referred to as tubes) guaranteeing the existence of viable solutions. Subsequently, we present a planning problem where the timing of vessel passages through designated waypoints within the previously identified tubes serves as decision variables.

We consider the scenario of the Guadalquivir River as an illustration, where navigation faces challenges due to its location within a highly protected natural area. This circumstance profoundly impacts the management of the Port of Seville,

© The Author(s), under exclusive license to Springer Nature Switzerland AG 2024   21
J. Moreno Nadales et al., *Optimal Vessel Planning in Natural Inland Waterways*,
SpringerBriefs in Applied Sciences and Technology,
https://doi.org/10.1007/978-3-031-64744-4_2

a critical intermodal logistics hub linking the Canary Islands with the mainland of Spain [1]. Given the limitations on making adaptations in such environmentally sensitive regions, the sole approach to enhancing waterway capacity involves optimizing vessel management. This necessitates the adoption of novel technologies and strategic management approaches within the framework of smart ports and Industry 4.0 [2]. The goal is to devise optimal voyage plans for a designated fleet of vessels arriving at both ends of the watercourse. These plans aim to minimize specific performance metrics dependent on vessel waiting and navigation times within the waterway, while adhering to various operational constraints. Each vessel's optimal voyage plan comprises a sequence of timings for crossing designated boundaries dividing the waterway into distinct sections, aiming to minimize performance indices and ensuring the existence of feasible trajectories adhering to all operational constraints for each vessel.

We illustrate that the proposed two-step approach is executable in real time using readily available solvers, even for significantly larger vessel volumes than those currently experienced at the Port of Seville. Furthermore, we validate its superiority over a commonly utilized method where vessel trip plans are determined based on a first-arrived first-served policy. This policy not only results in suboptimal decisions but also heightens the likelihood of accidents due to the human element involved in the process [3]. This concern will be extensively discussed in Chap. 3. It is essential to emphasize that the aim of this chapter is not to directly manage vessel movements within the waterway but rather to develop a planning tool that offers indicative timings for each vessel's passage across specific boundaries within the watercourse. Hence, the issue of vessel motion control is not tackled within this chapter, and it is assumed that the responsibility lies with the pilot to manage the vessel and adhere to the provided trip plan.

The theoretical contributions and results of this chapter were first presented in [4].

2.2 Problem Formulation

The objective of the waterway scheduling problem is to find the optimal trip plans for a set of N vessels that have to cross a natural waterway. We do not consider anchorage areas inside the waterway. Each of these vessels is characterized by its beam b_i, its maximum speed v_i, its draught δ_i, and the time when they arrive to the waterway and are ready to be scheduled, t_i^{ready}, where the integer $i \in I = \mathbf{N}_1^N = \{1, \ldots, N\}$ denotes the index of the vessel. We assume that vessels can navigate upstream and downstream. The set I_u includes all vessels sailing upstream, while the set I_d includes all vessels traveling downstream. For each vessel, the maximum trip duration is denoted as D_L.

We posit the existence of anchorage zones at both ends of the waterway, where vessels can await their scheduled entry into the watercourse. One approach to tackling the planning problem involves determining the velocity of each vessel through a discretization of time and space. However, this method typically yields exceedingly intricate problems. To circumvent this challenge, our proposed solution adopts the crossing times of designated boundaries as decision variables. For this purpose, we divide the waterway into Z sections, with the initial segment representing the river estuary and the concluding one denoting the entry lock of the inland port. Each section $p \in \mathbf{N}_1^Z$ is characterized by its length, d_p, the maximum permissible velocity for vessels within it, μ_p, and the effective width of the section, w_p. For safety considerations, the combined beams of any two vessels encountering each other are restricted by the minimum width of the section, ensuring safety during navigation.

Each vessel's itinerary is determined by the specific times it must traverse the boundaries between different sections, encompassing both ends of the watercourse, denoted by $p = 0$ and $p = Z$ respectively. To simplify notation, we represent these times as $t_{i,p}$, where in this context, $p \in \mathbf{N}_0^Z$ refers to the northernmost boundary of section p, including the extremes of the waterway situated in the river estuary and the entry lock of the port, designated as boundaries $p = 0$ and $p = Z$ respectively. The sequence of times $t_{i,p}$ increases as vessels travel upstream and decreases as vessels navigate downstream. By selecting these variables, all considered constraints can be effectively modeled, as outlined in the subsequent subsections. Knowing these times and the length of each section, river pilots can estimate the appropriate speed for vessels to traverse each section in accordance with the optimal plan.

In order to address this challenge while factoring in the varying depth of the waterway, a two-step methodology is introduced. Initially, the time windows within which each vessel can safely cross the waterway boundaries are determined, considering factors such as arrival times, maximum trip duration, and speed restrictions. An algorithm is proposed to compute these windows individually for each vessel, with the resulting sets referred to as safe tubes. The presence of a safe tube ensures that there exists at least one feasible trajectory satisfying all imposed constraints. Once the safe tubes for each vessel are established, a Mixed Integer Linear Programming (MILP) model is proposed to optimize the trip plans for all vessels, accounting for the fluctuating depth of the waterway. This model limits the sections of the channel where vessel encounters may occur based on vessel beams and waterway width. It is assumed that vessel encounters are managed in accordance with international regulations for collision prevention at sea [5] by river pilots, and that they do not affect the average sailing speed of the vessels.

Remark 2.1 Note that the aim of this chapter is not to directly regulate vessel movements, but rather to devise a tool for arranging vessel schedules in natural waterways and offering approximate safe crossing times for each boundary. Therefore, nautical phenomena such as the bank effect, shallow water effect, and vessel squat effect are not addressed in this text, as they do not impact scheduling. These effects are typically omitted in previous scheduling studies because they primarily influence motion control, a responsibility assumed to lie with the pilot in this chapter.

2.3 Safe Tubes Calculation Algorithm

To prevent grounding accidents, it is crucial to ensure that vessels are always located at time-space points where the depth of the waterway is greater than the draught of the vessels. Taking into account the time-varying depth of the waterway is a challenging task. To solve the planning problem, for each vessel i considered, we assume that there exists a set $\mathcal{K}_i$ of safe tubes. These safe tubes are regions in the time-space map that connect both extremes of the waterway, providing trip plans that satisfy depth and speed constraints. Each tube $k \in \mathcal{K}_i$ is defined by the minimum and maximum times, $t_{i,k,p}^{min}$ and $t_{i,k,p}^{max}$, respectively, in which vessel i has to cross each boundary p. Without loss of generality, we present next an algorithm that computes all the possible safe tubes corresponding to a given vessel traveling upstream. For vessels traveling downstream the procedure is equivalent.

Given the arrival time of vessel $i \in \mathcal{I}_u$ to the waterway, denoted as t_i^{ready}, along with the maximum allowable duration for the vessel's passage through the waterway, represented by D_L, our initial step involves computing, for each boundary $p \in \mathbf{N}_0^Z$, the set Q_p. This set encompasses all feasible time windows during which vessel i can traverse boundary p within the time interval from t_i^{ready} to $t_i^{ready} + D_L$. Essentially, Q_p consists of time intervals where the waterway's depth at the considered boundary, factoring in both the bathymetric profile and tide, is sufficient for the safe passage of the vessel. Typically, Q_p is constructed as a set of various non-overlapping time intervals denoted as Q_p^l, where $l \in \mathbf{N}_1^{card(Q_p)}$.

Figure 2.1 presents an explanatory instance of a waterway partitioned into three sections, where the vertical axis denotes time and the horizontal axis denotes space. Within the figure, the boxes at the boundaries depict the time intervals wherein the depth constraint is met. The shaded green intervals indicate the time windows deemed suitable for constructing the set of safe tubes.

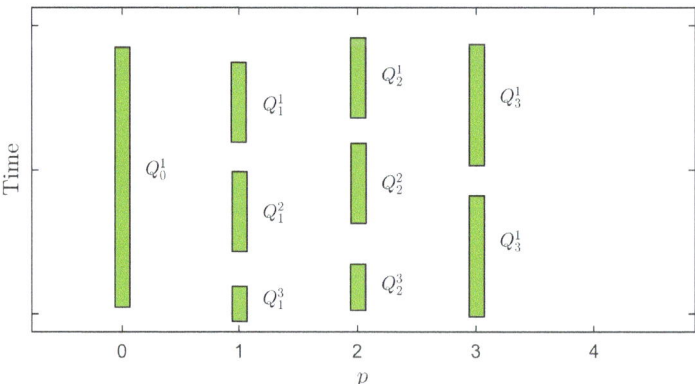

Fig. 2.1 Admissible crossing time windows Q_p of an example of a three-section waterway

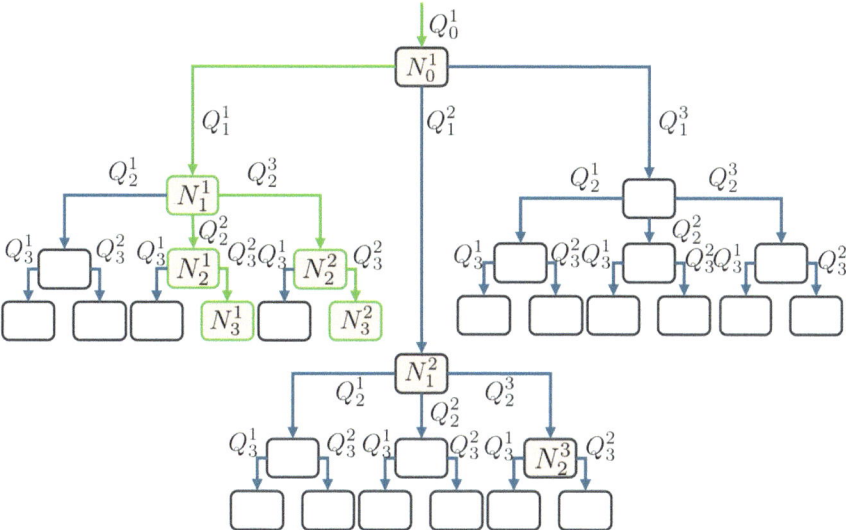

Fig. 2.2 Safe tube tree for the three-section example waterway

Once the set of permissible time windows for each boundary p have been identified, one might initially consider constructing the set of safe tubes by causal combinations of these windows. However, the potential number of safe tubes would grow exponentially with the number of sections. Figure 2.2 illustrates the tree structure generated for all conceivable sequences of time intervals across each of the four boundaries. This resulting tree comprises one level of depth for each boundary. It is noteworthy that the number of feasible sequences leading from boundary 0 to boundary 3 grows exponentially. In the present example, there exist $18 (1 \cdot 3 \cdot 3 \cdot 2)$ feasible sequences. Generally, this approach proves impractical. Furthermore, although any of these combinations would ensure compliance with depth constraints at the various waterway boundaries, uncertainty prevails regarding events while traversing each section, factoring in depth constraints and speed limits. To tackle this issue, we introduce the notion of a safe region between two-time windows belonging to consecutive boundaries and propose an algorithm to explore all potential combinations of time windows.

Definition 2.1 A safe region linking two-time windows is an area within the time-space map where it is feasible to discover at least one viable internal trajectory that does not violate either depth or speed limitations.

In Fig. 2.3, an example of safe region defined at section p, N_p^r, is displayed. The region is a trapezoid defined by four vertexes, A, B, C and D. In the figure, the dashed red lines represent trajectories described by vessels sailing at the highest allowed speed in the section, $\bar{\mu}_i(p)$, which is given by the minimum between the maximum speed allowed in section p, μ_p, and the maximum speed of vessel i, v_i.

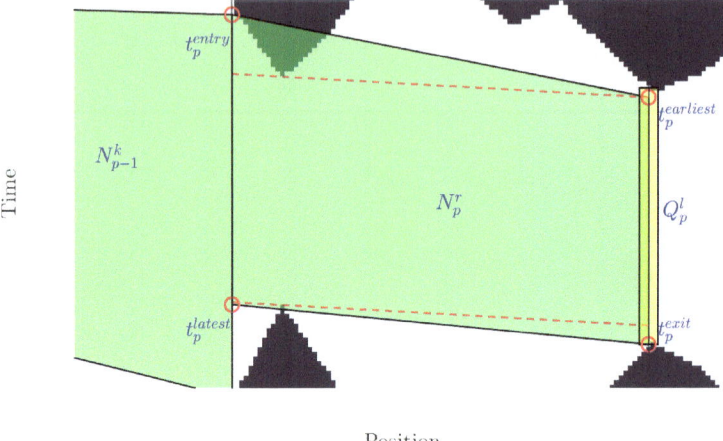

Fig. 2.3 Example of computing of a safe crossing region N_p^r connecting the previous one N_{p-1}^k to the border Q_p^l

Vertical lines correspond to a speed equal to zero, which implies that going from point A to point B (or from C to D) can be achieved by simply stopping the vessel at the corresponding boundary. Point A provides a bound on the earliest time (t_p^{entry}) when the vessel can cross safely the exit time window of the safe region defined in the previous section, N_{p-1}^k. Arriving earlier would imply abandoning the previous safe region N_{p-1}^k. Point B provides the latest time (t_p^{latest}) when the vessel can reach the exit time window of N_{p-1}^k. Arriving later at the left boundary would imply that the depth constraint would be violated, even if the vessel travels at the highest allowed speed (see the dashed line crossing through point B). If the vessel arrives at the boundary in the time interval $[t_p^{entry}, t_p^{latest}]$, then there exists at least one possible trip plan that can be followed without violating depth and speed constraints. Point C and D provide the earliest and latest time at which the vessel could safely get to the next border Q_p^l.

Various programming methodologies can be employed to determine the vertices A, B, C, and D defining the safe region N_p^r. In the simulation results presented in this section, point C is derived by identifying the highest point within boundary Q_p^l that can be reached from the left safe region N_{p-1}^k, traveling at the maximum permissible speed $\bar{\mu}_i(p)$, while adhering to depth constraints (i.e., avoiding entry into the dark blue region). It is worth noting that the slope of the dashed lines in the figure corresponds to the trajectory of a vessel traveling at speed $\bar{\mu}_i(p)$. Similarly, point B is determined by locating the lowest point on the left boundary, which belongs to the exit time window of the preceding safe region, and can be linked with the right time window Q_p^l by sailing at speed $\bar{\mu}_i(p)$ without encountering depth constraints. Point A represents the highest point of the left boundary within the previous safe region N_{p-1}^k. Lastly, point D denotes the lowest point on the border Q_p^l that does not breach depth constraints. This process must accommodate the time-space discretization of the predicted depth map $\mathcal{D}$. It is important to recognize that there may be instances where reaching Q_p^l from the previous safe region N_{p-1}^k without violating depth or speed constraints is impossible. In such cases, the corresponding safe region does not exist. The outlined procedure for determining each safe region is summarized in Algorithm (Fig. 2.3).

Algorithm 2.1 .

Require: $N_{p-1}^k, Q_p^l, \mu_p, d_p, i, v_i, \mathcal{D}$
1: $\bar{\mu}_i(p) = \min\{\mu_p, v_i\}$.
2: Find the earliest time $t_p^{earliest} \in Q_p^l$ when vessel i can exit safely section p starting from the exit time window of N_{p-1}^k. This time is obtained computing the highest point in Q_p^l that can be connected with the exit time window of N_{p-1}^k with a line of slope $\bar{\mu}_i(p)$ that does not violate depth constraints given by depth map $\mathcal{D}$.
3: Find the latest time $t_p^{latest} \in N_{p-1}^k$ when vessel i can enter safely section p exiting through Q_p^l. This is obtained computing the lowest point in the exit time window of N_{p-1}^k that can be connected with Q_p^l with a line of slope $\bar{\mu}_i(p)$ that does not violate depth constraints given by depth map $\mathcal{D}$.
4: If these times exist, then define N_p^r as the trapezoid given by vertexes A, B, C, and D, with C corresponding to $t_p^{earliest}$, B to t_p^{latest}, A to the highest point of the exit time window of N_{p-1}^k (t_p^{entry}), and D to the lowest point of Q_p^l (t_p^{exit}).

The aim is to establish safe tubes facilitating the secure traversal of the waterway by recursively linking safe regions between consecutive sections. This process can be conceptualized as a tree-building algorithm. During each iteration of this algorithm, this tries to concatenate the safe regions identified in the preceding iteration for the preceding waterway section p with the safe crossing windows delineated at the subsequent boundary of the waterway. At each step of the tree-building algorithm, given the exit time window of N_{p-1}^k and the time window to be reached, Q_p^l, it becomes essential to ascertain the existence of the corresponding safe region connecting both time intervals. Should such a connection be feasible, a new node is appended to the tree. This methodology is represented in the flow diagram illustrated in Fig. 2.4.

In Fig. 2.2, the resultant tree, depicted in green, corresponds to the waterway illustrated in Fig. 2.1. Within this figure, each labeled node denotes a safe crossing region facilitating the safe passage across the waterway up to the section represented by the respective level of the tree. Unlabeled nodes signify instances where it is not feasible to reach the subsequent time interval from the preceding safe region, thus requiring no further exploration. For instance, we posit that reaching Q_3^1 from N_2^1 is unattainable, implying the infeasibility of traversing Q_3^1 following the sequence Q_0^1-Q_1^1-Q_2^1-Q_3^1. In this scenario, neither Q_2^2 from N_1^2 nor Q_3^2 from N_3^3 are reachable, resulting in only two feasible safe tubes defined by the two children nodes of the tree, namely N_3^1 and N_3^2. The collection of safe tubes for a given vessel i, denoted as $\mathcal{K}_i$, is derived from the leaf nodes of the tree. For any leaf node k and the crossing interval at each boundary p-represented by the values of $t_{i,k,p}^{min}$ and $t_{i,k,p}^{max}$-these are determined

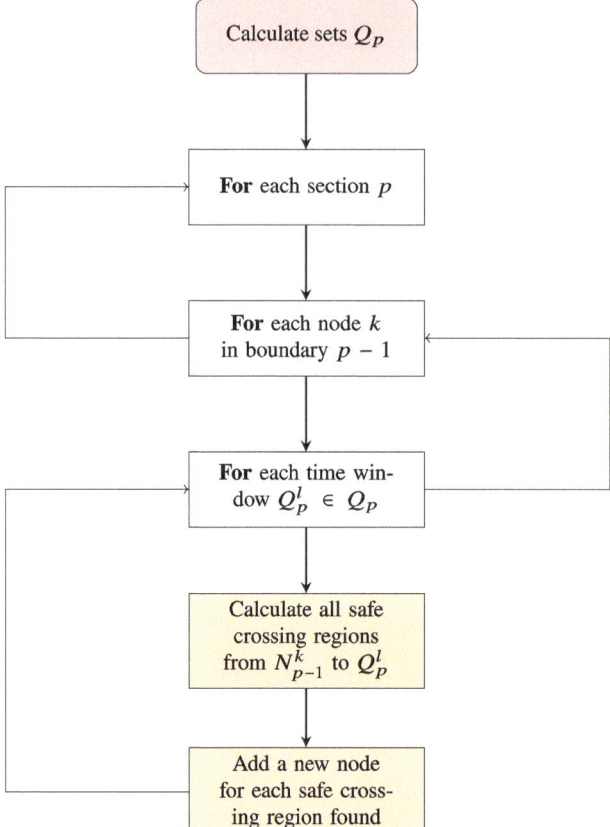

Fig. 2.4 Safe tube tree-building algorithm for vessels traveling upstream

by the entry times of the parent node at level p of the resultant tree, except for the time window at boundary Z, which is defined by the exit time window of the leaf node.

Remark 2.2 It is possible that the division of the river into sections leads to the emergence of enclosed areas in the depth map that are not intersected by any boundary. In such instances, the algorithm must be adjusted to address this issue. As illustrated in Fig. 2.5, the quantity of resulting safe crossing regions for each section would correspondingly increase in proportion to the number of enclosed regions not intersected by any boundary. In the example depicted in the figure, given the exit time window of the parent node N_p^k and the time interval to be reached, Q_p^l, two distinct nodes would be generated and incorporated into the tree.

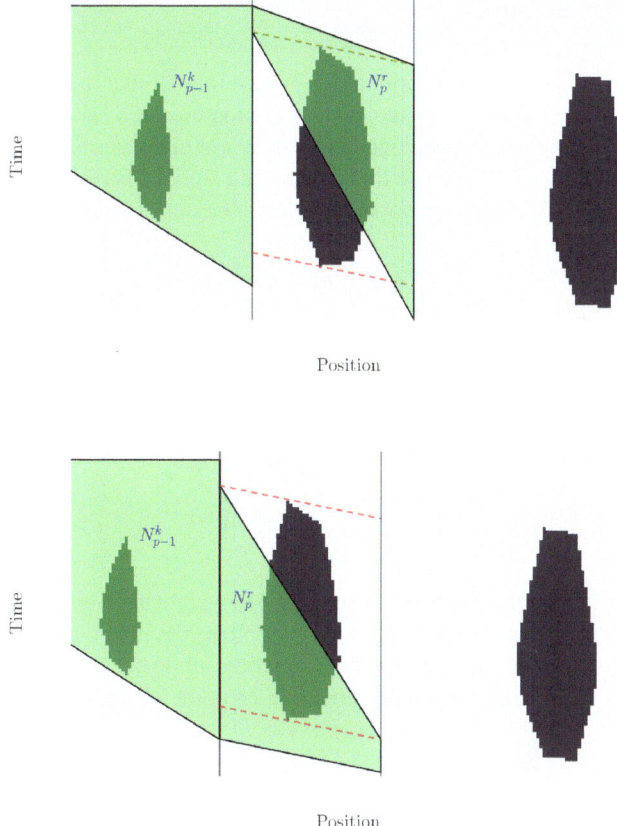

Fig. 2.5 Example of tubes in the case of existence of a closed region in the depth map that is not intersected by any boundary

2.4 Optimal Trip Planning

After identifying the safe tubes for each vessel, the scheduling problem for the vessels can be addressed. For this purpose, we introduce a formulation that adopts the structure of a MILP, encompassing both logic and dynamic constraints [6]. Below, we outline the cost function, constraints, and auxiliary decision variables necessary to characterize the problem.

2.4.1 Cost Function

The objective of the proposed formulation is to minimize the time vessels need to complete their trips along the waterway (navigation cost, C_N) and the time vessels spend waiting in the anchorage zone before they are given permission to enter the waterway (waiting cost, C_W). The time when each vessel i is ready to begin its trip is given by t_i^{ready}. Thus, the performance index to be optimized is the following:

$$J = \alpha \cdot C_N + \beta \cdot C_W, \tag{2.1}$$

with

$$C_N = \sum_{i \in \mathcal{I}_u} \left(t_{i,Z} - t_{i,0} \right) + \sum_{i \in \mathcal{I}_d} \left(t_{i,0} - t_{i,Z} \right) \tag{2.2}$$

and

$$C_W = \sum_{i \in \mathcal{I}_u} \left(t_{i,0} - t_i^{ready} \right) + \sum_{i \in \mathcal{I}_d} \left(t_{i,Z} - t_i^{ready} \right), \tag{2.3}$$

where α and β are parameters that can be chosen to weight differently navigation and waiting costs. Note that the navigation cost penalizes the time elapsed between the departure of each vessel and its arrival at destination, while the waiting cost penalizes the time elapsed between the vessel's arrival at its point of origin and its departure.

2.4.2 Speed Constraint

In each section of the waterway, there is a maximum speed that vessels cannot exceed given by $\bar{\mu}_i(p)$. This speed determines the minimum period of time needed for any vessel to sail through section p. To model this issue, for each section $p \in \mathbf{N}_1^Z$, and for each vessel $i \in \mathcal{I}_u$, the following constraints are included:

$$(t_{i,p} - t_{i,p-1}) \cdot \bar{\mu}_i(p) \geq d_p, \ t_{i,0} \geq t_i^{ready}. \tag{2.4}$$

In the same vein, for each section $p \in \mathbf{N}_1^Z$, and for each vessel $i \in \mathcal{I}_d$, the following constraints are included

$$(t_{i,p-1} - t_{i,p}) \cdot \bar{\mu}_i(p) \geq d_p, \ t_{i,Z} \geq t_i^{ready}. \tag{2.5}$$

2.4.3 Tube Constraints

To determine the safe tube for vessel $i \in \mathcal{I}$ to navigate among those identified using the algorithm detailed in the preceding section, denoted as $\mathcal{K}_i$, we introduce a set of binary decision variables $r_{i,k}$, where $k \in \mathcal{K}_i$. If $r_{i,k}$ equals 1, it indicates that vessel i must adhere to tube k for its journey. Conversely, all other variables corresponding to alternative safe tubes must be set to 0. To achieve this, the following set of constraints is established for each vessel i and for each $p \in \mathbf{N}_0^Z$:

$$t_{i,p} \leq t_{i,k,p}^{max} + M \cdot (1 - r_{i,k}), \tag{2.6}$$

$$t_{i,p} \geq t_{i,k,p}^{min} - M \cdot (1 - r_{i,k}), \tag{2.7}$$

$$\sum_{k \in \mathcal{K}_i} r_{i,k} = 1. \tag{2.8}$$

Given some tube k for some vessel i, we constraint the minimum and maximum times, $t_{i,k,p}^{min}$ and $t_{i,k,p}^{max}$, respectively, when the vessel can cross each section of the waterway according to the safe crossing windows that define the selected tube. Note that, by imposing (2.8), only a single tube is selected for each vessel.

2.4.4 Encountering Constraints

Finally, it is crucial to ensure that encounters between vessels, encompassing both overtaking and head-on maneuvers, occur at points with a practical width greater than the sum of the beams of the vessels involved. To achieve this, for each pair of vessels i and j with $i, j \in \mathcal{I}$ and $i < j$, a binary decision variable $c_{i,j,p}$ is introduced for every boundary $p \in \mathbf{N}_0^Z$ in the optimization problem. This variable is assigned a value of one if both vessels encounter each other head-on or overtake one another in section p. The value of the variables corresponding to forbidden maneuvres is set to zero, i.e.,

$$b_i + b_j > w_p \rightarrow c_{i,j,p} = 0 \quad \forall i, j, p. \tag{2.9}$$

We assume that each vessel i arrives at all boundaries of the waterway before vessel $j > i$ up to the boundary p where $c_{i,j,p}$ takes a value of 1. Following this occurrence, vessel j arrives at subsequent boundaries before vessel i. If the values of these variables for all sections p are equal to zero or if $c_{i,j,0} = 1$, it implies that vessels never encounter each other. Furthermore, a minimum time gap Δ_0 between two vessels crossing the same boundary is taken into account. To model these conditions, the following constraints are imposed:

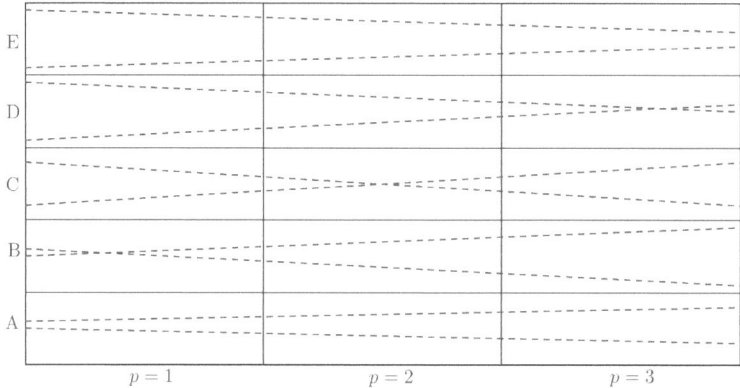

Fig. 2.6 Example of all possible crossing alternatives among two vessels sailing in opposite directions in a three-section waterways. **A** $c_{i,j,p} = \{1, 0, 0, 0\}$, **B** $c_{i,j,p} = \{0, 1, 0, 0\}$, **C** $c_{i,j,p} = \{0, 0, 1, 0\}$, **D** $c_{i,j,p} = \{0, 0, 0, 1\}$, **E** $c_{i,j,p} = \{0, 0, 0, 0\}$

$$t_{i,p} + \Delta_0 \le t_{j,p} + M \cdot \sum_{l \in \mathbf{N}_0^p} c_{i,j,l}, \qquad (2.10)$$

$$t_{j,p} + \Delta_0 \le t_{i,p} + M \cdot \left(1 - \sum_{l \in \mathbf{N}_0^p} c_{i,j,l}\right), \qquad (2.11)$$

where M is a number significantly larger than the value of $t_{i,p}$. To guarantee that each pair of vessels i, j can only encounter each other once during their trip, the following constraint is also included:

$$\sum_{p \in \mathbf{N}_0^Z} c_{i,j,p} \le 1. \qquad (2.12)$$

In Fig. 2.6, all potential encountering scenarios for two vessels sailing in opposite directions are depicted within a simple example of a waterway divided into three sections. The horizontal axis represents the distance between both ends of the waterway, with the boundaries of the sections indicated by continuous blue lines. Time is represented on the vertical axis, with the initial time positioned at the top of the figure. Vessel trajectories are illustrated by dashed blue and red lines. A similar figure can be generated for vessels traveling in the same direction.

Remark 2.3 The encountering situations constraint can be modified to allow several overtaking or head-on maneuvers among two vessels.

Remark 2.4 It is assumed that vessels undergoing any encountering situation do not change their speed during the maneuver.

2.4.5 Optimal Scheduling Problem

The optimization problem proposed to find the optimal values of the decision variables that define the optimal sailing plan, t^*, c^* and r^* can be formulated as

$$\min_{t^*,c^*,r^*} J \text{ s.t } (4) - (12). \tag{2.13}$$

In the proposed optimization problem, we assume that the primary goal is to determine the trip plans for all vessels under consideration. However, it is also feasible to account for vessels with fixed trip plans, thus accommodating various scenarios encountered during normal operational conditions. These scenarios include situations such as integrating a newly arrived vessel into a waterway already occupied by vessels in transit or adjusting schedules in response to unforeseen circumstances. It is noteworthy that the arrival time of any vessel could be enforced by introducing an additional constraint that sets the value of $t_{i,Z}$ or $t_{i,0}$ to the desired time for vessels traveling upstream and downstream, respectively.

Remark 2.5 The proposed method is based on resolving an NP-hard optimization problem, rendering it possibly impractical for situations involving a considerable number of vessels or waterway sections. Nonetheless, in the simulation section, we present evidence that the proposed solution is viable for the present requirements of the Port of Seville. The complexity of the optimization problem predominantly depends on three factors: the volume of vessels N, the quantity of sections Z, and the number of safe tubes for each vessel. Specifically, the number of constraints and binary variables increases quadratically with N and linearly with Z and the number of tubes.

2.5 Case Study

To show the benefits of the scheduling methodology presented in this chapter, we apply the proposed trip planning approach to the Guadalquivir River, a natural inland waterway located in the south of Spain, connecting the inland Port of Seville to the Atlantic Ocean. To verify the effectiveness of the proposed method, various randomly generated scenarios are considered. All optimization problems are solved using the Gurobi solver [7]. The simulations are conducted on an AMD Ryzen 5 platform running at a speed of 3 GHz.

The Guadalquivir River is segmented into a total of $Z = 24$ sections, resulting in 25 boundaries in total. Each segment p is distinguished by its length d_p, the maximum sailing speed for vessels μ_p, and its width w_p, as detailed in Table 2.1.[1] Vessels' beams are categorized into three groups: small, medium, and large, represented by

---

[1] All data used to construct various scenarios are provided by local authorities with slight modifications for confidentiality reasons.

Table 2.1 Guadalquivir waterway parameters

Id	d (km)	μ (km/h)	w	Id	d (km)	μ (km/h)	w
1	1.9	10.1860	1	13	4.2	19.4460	8
2	1.4	16.6680	4	14	4.5	19.4460	1
3	4.5	16.6680	2	15	3.5	22.2240	8
4	4.2	17.5940	4	16	1.2	18.5200	4
5	2.0	17.5940	4	17	3.4	21.2980	4
6	5.2	18.5200	4	18	1.7	23.1500	4
7	6.6	21.2980	4	19	2.0	23.1500	4
8	3.5	21.2980	4	20	4.2	23.1500	4
9	5.9	21.2980	2	21	3.0	22.2240	1
10	5.4	22.2240	4	22	2.2	18.5200	8
11	2.1	22.2240	4	23	4.8	21.2980	1
12	4.7	22.2240	1	24	4.9	21.2980	2

values of 1, 2, and 4, respectively. The width of each section is assigned a value from the discrete set {2, 4, 8} instead of its actual measurement. Vessels can encounter each other in a section only if the sum of their beams does not exceed the width of the waterway in that segment.

In all the simulations carried out in this section, the objective is to minimize the total trip time of all vessels considered, and hence, the weights α and β are both set to 1. In addition, we consider a minimum time between crossings of $\Delta_0 = 1$ h/10.

To consider the influence of tides, we have taken into account the depth map shown in Fig. 2.7. This map has been created by overlaying the bathymetric data of the river with an interpolated tide prediction map derived from predictive models incorporating sensory inputs. By utilizing this depth map, it becomes possible to identify the time-space coordinates where a given vessel can navigate while accounting for its draft. Subsequently, this map can be utilized to determine the set of safe tubes. The map depicted in Fig. 2.8 has been generated for vessels with a draft of 7.5 m, with the dark blue areas indicating regions that vessels should avoid to prevent grounding incidents.

In order to analyze the performance of the proposed approach, it has been applied to a number of different randomly generated scheduling scenarios. Each scenario is defined by the following parameters:

- N: number of vessels considered, half of them sailing upstream and the other half sailing downstream.
- t^{init}: Initial time of the scenario.
- λ: Characteristic parameter of a Poisson distribution used to define the time interval between two consecutive arrivals of vessels to the same end of the waterway.
- D_L: Maximum period of time a vessel is allowed to be sailing before reaching its final destination.

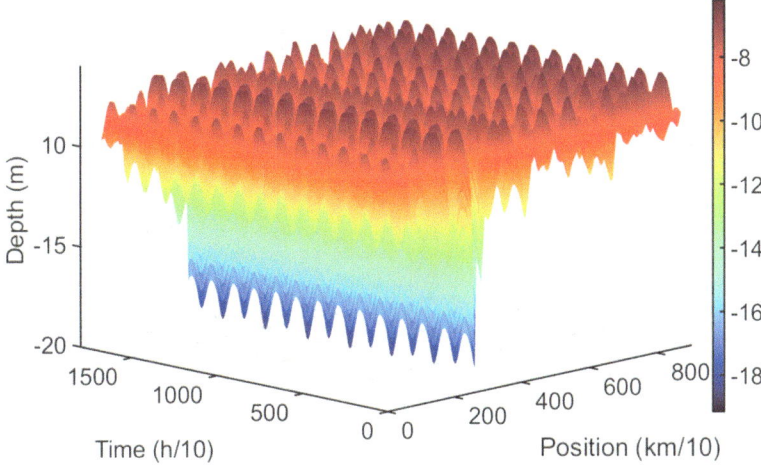

Fig. 2.7 Prediction of the depth of the Guadalquivir waterway for the week considered

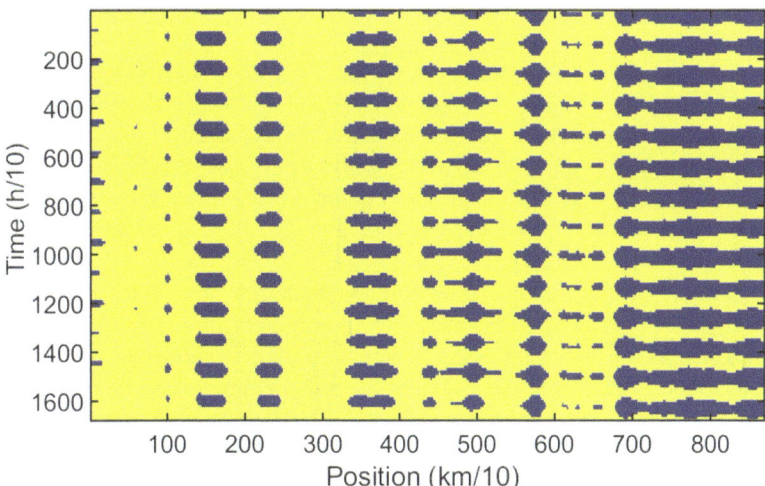

Fig. 2.8 Map of time-space regions in which a 7.5 m draught vessel may sail for the week considered

Each scenario generated using these parameters is defined by the sequence of times at which vessels arrive at the waterway. The sequence at each end of the waterway follows a Poisson distribution with a characteristic parameter λ, with both sequences starting at time t^{init}. The width of each vessel is randomly determined from a uniform distribution. We assume that all vessels have a draft of 7.5 m, approximately corresponding to the draft of the largest vessel currently operating in the Port of Seville, to address the worst-case scenario.

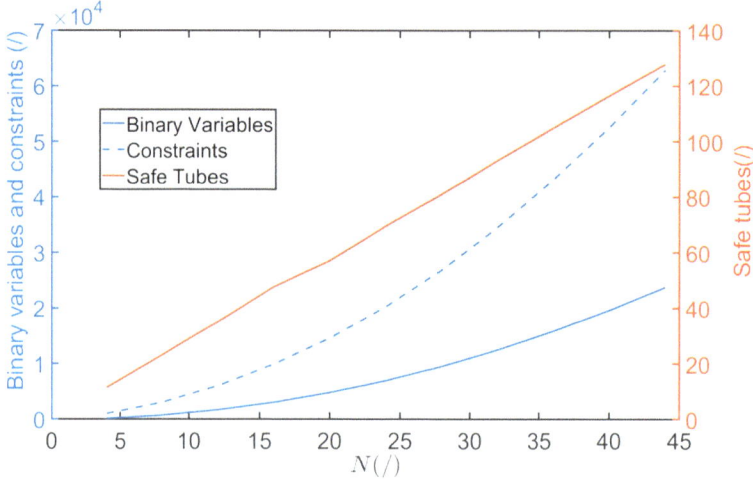

Fig. 2.9 Number of binary variables, constraints, and safe tubes as a function of N

We show the simulations carried out to test the proposed approach and analyze the obtained results. We show the main properties of the proposed approach in terms of computational complexity, optimality, and feasibility, and how this approach out-performs a first-arrived first-served policy which optimizes vessels as they arrive to the waterway without modifying the trips of other vessels already sailing through it. Due to the discretization of the map of depth, units h/10 and km/10 are employed in all the simulation.

The complexity of the optimization problem predominantly hinges on the number of vessels N. We proceed to illustrate how the number of constraints and binary variables increases with the number of vessels N, sections Z, and tubes. Additionally, we analyze the impact of these variables on computational time. Since the MILP problem is NP-hard, the computation time exhibits exponential growth relative to the number of binary variables.

Figures 2.9 and 2.10 show the effect of the number of vessels considered, N, on the complexity of the problem, the different costs, and the computing time. The results shown are the average values of 50 different scenarios for increasing values of N and fixed values of λ and D_L. The initial time of each scenario is chosen randomly in the first 24 h of the week. In this set of scenarios, vessels arrive both at the port of Seville and the river estuary with a mean frequency of $\lambda = 3$ h and have a maximum of $D_L = 18$ h to reach their destination. In these simulations, the width of the different vessels has been randomly generated.

Figure 2.9 shows the number of binary variables, constraints, and safe tubes. This figure shows that the number of binary variables and constraints grow polynomially with N, while the number of safe tubes is almost linear.

Figure 2.10 shows the computation time needed to solve the problem using Gurobi, and the waiting and navigation costs. This figure shows that the computation time

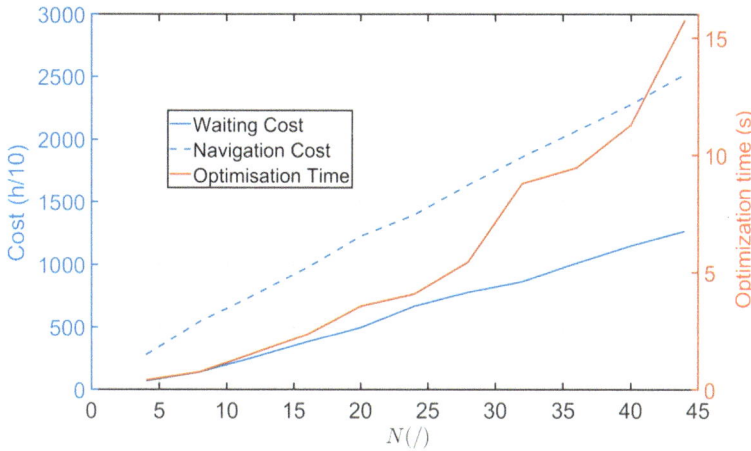

Fig. 2.10 Optimization time; waiting and navigation costs

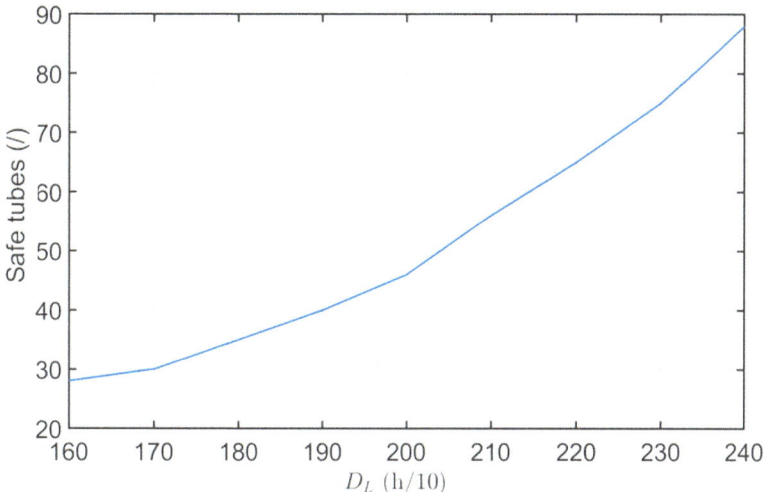

Fig. 2.11 Number of safe tubes with respect of D_L for a fixed value of t_i^{ready}

grows exponentially as expected while the costs grow almost linearly. We notice that, even for a large number of vessels, the optimization problem is solved within a reasonable time. Note that, because the set of safe tubes can be computed in advance knowing the prediction of the time, this time has not been considered.

Figure 2.11 shows the effect of the maximum trip time D_L on the complexity of the problem in terms of the number of safe tubes for a total of 12 vessels, i.e., $N = 12$ and a fixed value of $\lambda = 3$ h. As it can be seen, the number of safe tubes grows with D_L. In Table 2.2 we show the time it takes to compute these sets of safe tubes applying the procedure detailed by the flow diagram depicted in Fig. 2.4 as a

Table 2.2 Safe tube computing time as a function of D_L

D_L [h/10]	100	120	140	160	180	200
Comp. time (s)	1.48	1.78	2.14	2.57	3.25	4.22

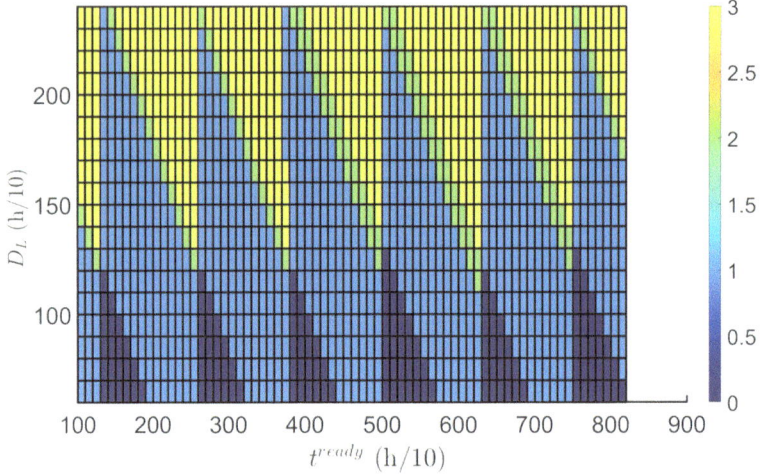

Fig. 2.12 Number for tubes generated for vessels sailing upstream as a function of D_L and t_i^{ready}

function of D_L. As it can be appreciated, computing time grows exponentially with the value of D_L. It is also interesting to see how, given the periodic nature of the tide, the number of tubes generated strongly depends on t_i^{ready}. This is shown in Figs. 2.12 and 2.13 for vessels sailing upstream and downstream, respectively. As it can be seen, both grow exponentially with D_L, but the number of tubes generated when sailing downstream is greater than the number of tubes generated when sailing upstream. This is because in the case of vessels sailing upstream, the only option is to surf the wave generated by the tide, while when sailing downstream the vessel may have to wait for the next tide to overcome some points of the waterway, and often this can be done in different times of the trip, leading to multiple options to arrive before the time limit $t_i^{ready} + D_L$.

Concerning waiting and navigation expenses, it is essential to note that these can be influenced by the values of the waiting parameter α and the navigation parameter β. To observe the impact of these parameters on the costs included in the performance index, simulations were conducted by varying the value of α while keeping β constant. These simulations were performed with D_L fixed at 200 h/10, considering $N = 6$ vessels, and with the arrival frequency parameter set at $\lambda = 3$ h. The obtained results represent the mean values across a series of 100 randomly generated scenarios, as depicted in Table 2.3.

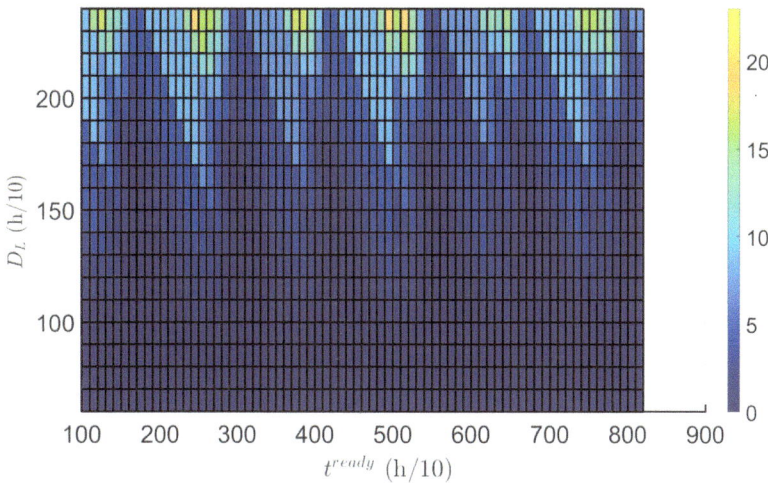

Fig. 2.13 Number for tubes generated for vessels sailing downstream as a function of D_L and t_i^{ready}

Table 2.3 Effect of the weighting parameters on waiting and navigation costs

α	β	C_W	C_N
0.5	1	21.13	686.87
1	1	76.36	631.5
1.5	1	319.24	388.62
2	1	533.26	275.61
2.5	1	637.25	223.61

Choosing an appropriate value for D_L is important because if it is chosen too small, it may not be possible to reach the destination during that period, yielding and infeasible problem. To demonstrate this issue, we present a simulation in which 50 random scenarios are generated fixing $N = 12$ and $\lambda = 3$ h for decreasing values of D_L. The results are shown in Table 2.4. For each value of D_L, the percentage of scenarios starting in an initial random time on the first 24 h which are not feasible (% UF) are obtained. As expected, if D_L is too small, the vessels cannot reach their destination because no feasible tube can be found. It is important to remark that those vessels that need longer trip times are those traveling against the tide from the estuary of the river to the Port of Seville. These vessels need over 16 h to ensure that a feasible tube will exist. Remember that the worst-case scenario is considered, i.e., vessels of a draught equal to 7.5 m.

We present next some experiments that demonstrate that using the proposed planning approach can improve the performance of the port. To this end, the proposed approach is compared to the baseline solution in which in each scenario, vessels are

Table 2.4 Effect of D_L on the feasibility of the problem

D_L [h/10]	150	160	170	180	190	200	210	220
% UF	100	86	50	6	0	0	0	0

Table 2.5 Effect of λ on the feasibility of the problem

λ [h/10]	1	3	5	10	30	60	120
Large size							
% UF baseline	100	100	100	100	100	0	0
% UF proposed	100	31	0	0	0	0	0
Random size							
% UF baseline	36	35	34	28	27	2	0
% UF proposed	0	0	0	0	0	0	0

scheduled individually as they arrive to the waterway. This is a first-arrived first-served (FA/FS) approach in which an operator schedules each vessel optimally considering the speed, depth, and width constraints while fixing the plan of the previous vessels. This first-arrived first-served approach does not profit from the possibility of planning all vessels simultaneously, leading to a worse performance as demonstrated in the results. The planning in the first-arrived first-served approach is solved by fixing the values of the previously planned trips for vessels getting earlier to the waterway.

The first simulation demonstrates that, for an increasing frequency of vessel arrivals (λ), feasibility issues may arise if vessels are scheduled individually, which in real operation means that vessels inside the waterway may have to modify their previously decided plans. The results are shown in Table 2.5, where the percentage of unfeasible scenarios using both the baseline and the proposed approach are shown (UF Baseline and UF Proposed, respectively). In these scenarios, $N = 8$, $D_L = 18\,\text{h}$ and $t^{init} = 0$. We consider two different cases. In the first one, to account for the worst-case scenario, all vessels are large, i.e., $b_i = 4$. In the second one, the size of the vessels is randomly selected. For each value of λ, 50 scenarios are generated randomly. It can be seen how the proposed approach is always better (not worse) than the first-arrived first-served approach.

After showing that the proposed approach outperforms the first-arrived first-served one in terms of feasibility, we show that the cost of the proposed solution is always better than the cost of the first-arrived first-served solution, thus reducing the total time vessels spend waiting and navigating through the waterway. Again, 50 random scenarios for $N = 8$ are generated fixing the initial time to some random value in the

Table 2.6 Effect of λ on the cost of the problem comparing the costs of the proposed solution with those obtained when the first-arrived first-served (FA/FS) approach is adopted

λ [h/10]	C_W FA/FS	C_W	C_N FA/FS	C_N	ΔC
3	201.45	160.99	615.33	612.94	42.84
5	230.34	188.18	617.55	624.86	34.84
10	245.12	208.93	595.34	599.27	32.26
30	240.41	206.27	577.58	582.21	29.51
60	200.84	253.62	628.23	555.83	19.62
120	50.32	155.93	733.98	628.37	0

first 24 h of the week, D_L to 18 h and increasing the value of λ from 6 min to 1 h. In this case, the size of all vessels is randomly selected. The results obtained are shown in Table 2.6. As it can be seen, the proposed solution outperforms the first-arrived first-served one in terms of cost and the difference between the cost of the first-arrived first-served solution and the proposed solution, ΔC, grows as λ decreases.

2.6   Conclusions

In this chapter, we proposed a two-step optimization-based approach to schedule and plan vessels in natural waterways with a time-varying depth, taking the time vessels cross through to each boundary and dividing the waterway as decision variables. An algorithm is proposed to find the optimal windows in which vessels should cross each delimiting point to satisfy the depth constraint. We demonstrated by extensive simulation of different scenarios in the Guadalquivir River that the proposed formulation can be applied to solve the planning problem for a set of vessels and that the computational time is tractable for the current demands of the Guadalquivir River. It is shown that the solution of the proposed problem improves the fist-arrived first-served solution in terms of optimality and feasibility.

References

1. J. Muñuzuri, L. Onieva, P. Cortés, J. Guadix, Using IoT data and applications to improve port-based intermodal supply chains. Comput. & Ind. Eng. **139**, 105668 (2020)
2. I. de la Peña Zarzuelo, M.J.F. Soeane, B.L. Bermúdez, Industry 4.0 in the port and maritime industry: a literature review. J. Ind. Inf. Integr. 100173 (2020)
3. C. Chauvin, S. Lardjane, G. Morel, J.P. Clostermann, B. Langard, Human and organisational factors in maritime accidents: analysis of collisions at sea using the HFACS. Accid. Anal. & Prev. **59**, 26–37 (2013)
4. J. Nadales, D.M. De La Peña, D. Limon, T. Alamo, Safe navigation through waterways of time-varying depth based on reachability analysis, in *2023 European Control Conference (ECC)*. (IEEE, 2023), pp. 1–6

5. I.M. Organization, COLREG: convention on the international regulations for preventing collisions at sea, 1972. International Maritime Organization (2002)
6. A. Bemporad, M. Morari, Control of systems integrating logic, dynamics, and constraints. Auto-
   matica **35**(3), 407–427 (1999)
7. Gurobi Optimization, LLC: Gurobi Optimizer Reference Manual (2021). https://www.gurobi.
   com

Chapter 3
Safety-Oriented Rescheduling in Inland Waterways

Abstract This research introduces a framework designed to identify, classify, and effectively reschedule major accidents occurring in natural inland waterways. Leveraging real-time vessel positioning data from monitoring systems, deviations from planned trajectories are detected, and a classification algorithm determines the nature of each incident. To facilitate optimal rescheduling, diverse strategies tailored to the specifics of each incident and the typical challenges of these waterways are proposed. The overarching goal is to minimize disruptions and limit the number of affected vessels, thereby safeguarding the port's reputation. To validate the efficacy of the methodology, a practical application is conducted in the Guadalquivir River, with a comparative analysis of the outcomes achieved through various rescheduling strategies.

3.1 Introduction

In this chapter we develop a monitoring and rescheduling solution for vessels in natural inland waterways. When an anomaly is detected and identified, vessel itineraries undergo rescheduling through the solution of a modified optimization problem based on the work developed in Chap. 2. The proposed rescheduling method takes into account operational objectives, such as sailing times, while factoring in the impact of the identified incident on the original schedule, employing a multi-objective optimization approach. These objectives frequently conflict with each other since optimizing the performance of all vessels under the new conditions might incur additional costs due to deviations from the initial trip plans. To address this issue, a weighted sum approach is employed [1], allowing the port operator to generate different plans by adjusting a single weighting parameter.

We introduce a novel monitoring and rescheduling approach designed for vessels navigating natural inland waterways. Our novel approach is tailored to address diverse scenarios in which vessels encounter unexpected events, such as arrival delays or machinery failures, which may necessitate alterations to their original voyage plans. A comprehensive monitoring system is deployed to swiftly detect incidents, while a decision tree classifier is utilized to categorize the nature of each incident.

© The Author(s), under exclusive license to Springer Nature Switzerland AG 2024
J. Moreno Nadales et al., *Optimal Vessel Planning in Natural Inland Waterways*,
SpringerBriefs in Applied Sciences and Technology,
https://doi.org/10.1007/978-3-031-64744-4_3

Upon detecting an incident, our system calculates a new optimal trip plan for all vessels, factoring in the specific type of incident identified and a range of distinct rescheduling criteria. To evaluate the effectiveness of our proposed approach, we conducted simulations across various scenarios. Specifically, we scrutinized how our methodology performs when vessels confront arrival delays, experience speed reductions, or encounter machinery failures necessitating repairs. The outcomes of these simulations affirm the effectiveness of our approach in managing such scenarios. It excels at detecting irregularities in vessel trajectories and promptly adapting voyage plans in real time, employing a multi-objective optimization framework that simultaneously accounts for the impact of the incident on the original schedule and optimizes vessel performance.

The theoretical contributions and results of this chapter were first presented in [2].

3.2 Problem Formulation

We consider a fleet consisting of N vessels operating within a bidirectional natural watercourse, linking an inland cargo terminal to either the open sea or another inland terminal. These vessels possess the ability to navigate both upstream and downstream. Our premise for ensuring the safe and effective navigation of these vessels revolves around the involvement of experienced river pilots tasked with regulating vessel speeds in accordance with an optimized trip plan. This plan is crafted to minimize vessel operational expenses while complying with a series of security and operational restrictions.

The trip plan consists of a series of timing points at which each vessel should pass through specific waypoints, effectively dividing the waterway into Z distinct sections. For each vessel indexed as $i \in I = \mathbf{N}_1^N$, where $\mathbf{N}_1^N$ represents the set of natural numbers from 1 to N, this trip plan is denoted as $\mathbf{t}_i \in \mathbb{R}^{Z+1}$, with $t_{i,p}$ for all $p \in \mathbf{N}_0^Z$ being the time when vessel i crosses boundary p. Assuming that vessels maintain a constant speed within each section, this plan defines the expected trajectory of each vessel, denoted as $z_i(t)$.

Unexpected events can cause vessels to veer off their intended routes, presenting risks to operational efficiency and safety. To address this challenge, our research presents a continuous monitoring system tailored to monitor vessels' compliance with their planned paths. This system is specifically designed to detect any deviations from the expected trajectory and promptly alert port authorities, empowering them to take proactive steps to prevent potential accidents and safeguard ongoing operations. By implementing this monitoring system, we bolster the overall safety and dependability of the operation, guaranteeing that vessels can navigate the waterway securely and efficiently.

The fundamental principle underlying the monitoring system revolves around real-time comparisons between each vessel's current location, as provided by the AIS, and its anticipated location. If any vessel deviates from its expected trajectory by a predetermined threshold, a warning event is triggered, indicating the detection of an incident requiring rescheduling to establish a new trip plan, with a priority

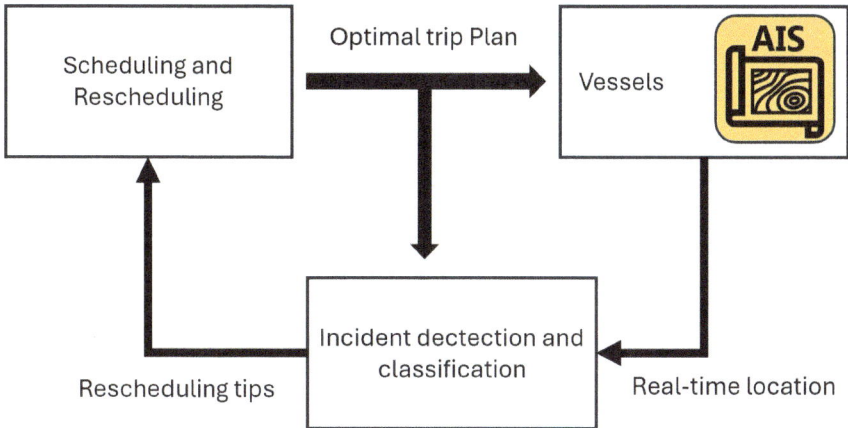

Fig. 3.1 Architecture of the vessel monitoring, detection and classification, and rescheduling system

on operational safety. Upon detection, the incident is subsequently classified into one of the predefined categories. It is noteworthy that the timing of rescheduling varies depending on the type of incident. For instance, in the case of breakdown incidents, rescheduling is scheduled upon vessel repair, while in instances of speed reduction incidents, it is initiated immediately upon detection. Furthermore, distinct rescheduling guidelines must be employed contingent upon the nature of the incident. Consider, for example, a scenario involving a speed reduction incident, in which case the new maximum vessel speed must be updated before rescheduling can commence.

We provide a comprehensive overview of the array of rescheduling strategies available to rectify deviations in vessel trajectories. These strategies encompass optimizing sailing times and minimizing the impact of the identified anomaly on other vessels. Upon computation of the new optimal trip plan based on the chosen rescheduling strategy, it is promptly communicated to the vessels to ensure they are informed about the alterations to their trajectories. The proposed monitoring and rescheduling approach, depicted in Fig. 3.1, serves as a crucial mechanism for enhancing both the efficiency and safety of overall vessel operations through continuous monitoring and dynamic rescheduling. In the diagram, the monitoring system receives as input the real-time position of each vessel and the expected position according to the planned trajectory. The output comprises rescheduling guidelines, which are then fed into the rescheduling module.

3.3   Incident Detection and Identification

Inland waterways are prone to various incidents that pose significant risks to vessel safety. These incidents encompass a wide range of factors, each presenting its own set of associated risks and challenges. For example, delays can stem from various sources within an inland waterway, such as heavy traffic, navigational challenges, obstacles like debris, or inaccuracies in estimated arrival times. These delays not only disrupt the seamless flow of vessel traffic but also contribute to congestion, increasing the probability of collisions or other accidents.

Mechanical failures pose another significant incident that can jeopardize vessel safety. These failures can encompass a variety of issues, ranging from engine malfunctions to steering system problems and breakdowns in propulsion equipment. When a vessel encounters a mechanical failure, it may become immobilized, lose control, or drift into restricted areas, thereby increasing the risk of accidents.

Moreover, unforeseen maintenance issues may emerge within inland waterways, demanding immediate attention and potentially disrupting vessel operations. The severity of these maintenance issues can vary, spanning from minor repairs to more extensive tasks necessitating prolonged periods of vessel downtime. Unpredicted maintenance challenges not only contribute to delays but can also worsen congestion or operational disruptions, thereby heightening the risk of accidents.

Our focus is on three specific categories of occurrences:

- **Arrival delay**: This happens when a vessel reaches the waterway later than initially expected in the original timetable [3].
- **Breakdown**: This situation arises when a vessel encounters a mechanical malfunction, leading to a cessation of its navigation.
- **Speed reduction**: In this circumstance, a vessel must decrease its velocity, often as a precautionary measure to prevent engine overheating. Additionally, various factors such as weather conditions, water currents, and vessel dimensions can influence the vessel's pace [4].

Inland waterways typically encounter these three types of incidents, each requiring specific attention from port authorities. Addressing these incidents is time-intensive and may arise at any hour of the day. Our goal is to enhance vessel safety, improve operational efficiency, and reduce risks in inland waterway navigation by understanding the underlying causes and implementing appropriate mitigation measures for these incidents [5].

We present an algorithm for detecting and identifying anomalies based on the AIS signal. In the previous chapter, we introduced a scheduling algorithm aimed at determining the optimal crossing times for vessels at each boundary within the waterway, denoted as $t_{i,p}$. These times provide vessels with an optimal trip plan, represented as $\mathbf{t}_i$. Assuming vessels maintain a constant speed within each section of the waterway, it becomes feasible to compute the expected location of each vessel, denoted as $z_i(t)$. To ensure vessel compliance with their optimal trip plans, we utilize a monitoring system. This system continuously compares the expected vessel location, $z_i(t)$, with the actual location provided by the AIS, denoted as $z_i^{\dagger}(t)$. If a significant deviation exists between the expected and actual locations, surpassing a predefined threshold value ϵ_z, an unexpected incident is flagged. The threshold value ϵ_z represents the maximum allowable deviation between the expected and actual vessel locations. By continuously monitoring vessels, this system can issue early warnings of potential incidents, allowing authorities to intervene before accidents occur.

Let $t^{current}$ denote the current time. Let $\bar{v}_i$ represent the average speed of vessel i calculated based on the last N_w measurements from the AIS signal. v^{break} denotes the breakdown threshold set to identify a breakdown incident. We assume that no accidents occur until rescheduling is carried out after an active incident. If an incident is detected, it is categorized as an *arrival delay* if the vessel is positioned at one of the waterway's extremes. If the vessel is navigating through the waterway at the time of the incident's detection and the estimated average speed $\bar{v}_i$ falls below the breakdown threshold v^{break}, the incident is classified as a *breakdown*. Otherwise, it is classified as a *speed reduction*. This classification process is represented by the decision tree depicted in Fig. 3.2, illustrating the comparison between expected and actual positions to determine the incident type.

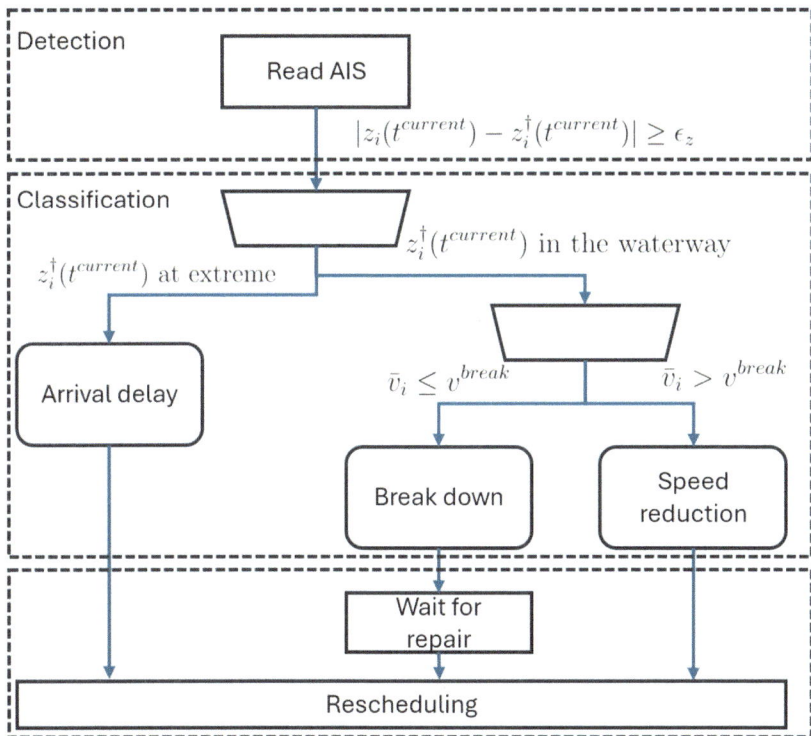

Fig. 3.2 Decision tree for classifying vessel incidents. In this classification tree, $z_i(t^{current})$ refers to the expected location of vessel i at current time $t^{current}$ according to the optimal plan. $z_i^{\dagger}(t^{current})$ refers to the location of vessel i provided by the AIS signal also at current time $t^{current}$. $\bar{v}_i$ is the average speed of vessel i estimated using a certain number of past location data from the AIS signal. v^{break} is the threshold for determining break down incidents

3.4   Rescheduling Strategies

In the event of unforeseen incidents, such as those previously described, occurring in inland waterways, it may be necessary to reschedule vessel trips to ensure safe and optimal operations under the new circumstances. This involves analyzing the updated conditions and adjusting vessel schedules, considering factors such as the incident's nature and severity, existing traffic conditions, and relevant regulations or guidelines. Suppose an unexpected incident involving vessel i^* is detected at time $t^\times$. At that moment, the location of each vessel i already navigating in the waterway is denoted as $z_i^\dagger(t^\times)$, and the last boundary it crossed before the incident is identified as $p_i^\times$. To determine the new optimal trip plan $\mathbf{t}$ for all vessels, a fresh optimization problem must be solved, integrating the current location of all vessels already underway in the waterway.

In case of a speed reduction incident, the rescheduling process involves adjusting the maximum speed at which the affected vessel can sail to $\bar{v}_{i^*}$. This speed is estimated based on the last N_A AIS signals. Rescheduling occurs immediately upon detecting the incident at time $t^\times$. For breakdown incidents, the rescheduling problem is addressed after the affected vessel is ready to resume its journey, denoted as t^{repair}. In cases of arrival delay incidents, rescheduling begins after the vessel has arrived at the waterway and the incident has been detected. The timing of rescheduling varies depending on the incident type. Figures 3.3, 3.4 and 3.5 provide a graphical representation of when each incident is detected based on its type. Arrival delay incidents are detected when the vessel reaches one of the waterway's extremes. Speed reduction incidents are detected sometime after the speed reduction. Breakdown incidents are detected after the vessel breaks down and stops sailing.

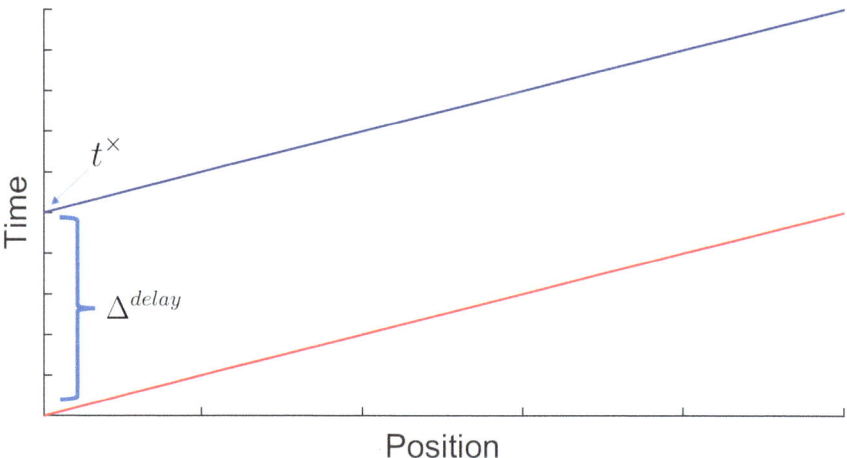

Fig. 3.3 Detection and rescheduling time for arrival delay incidents. Δ^{delay} is the time difference between the expected and real arrival times to the waterway

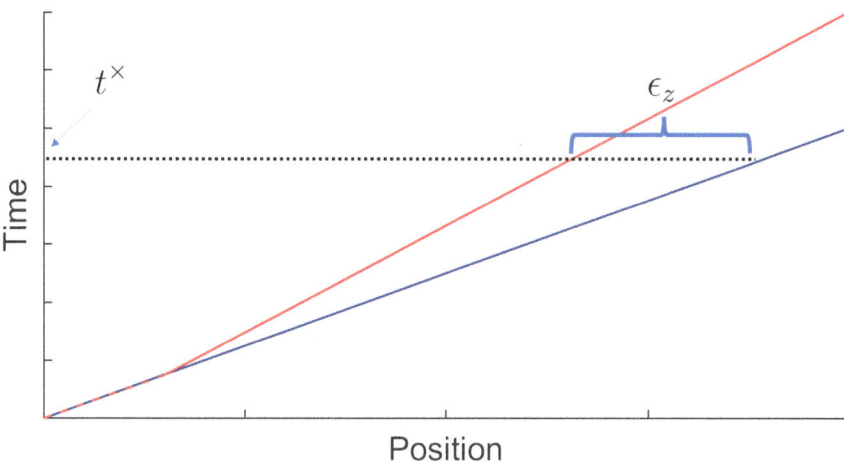

Fig. 3.4 Detection and rescheduling time for speed reduction incidents

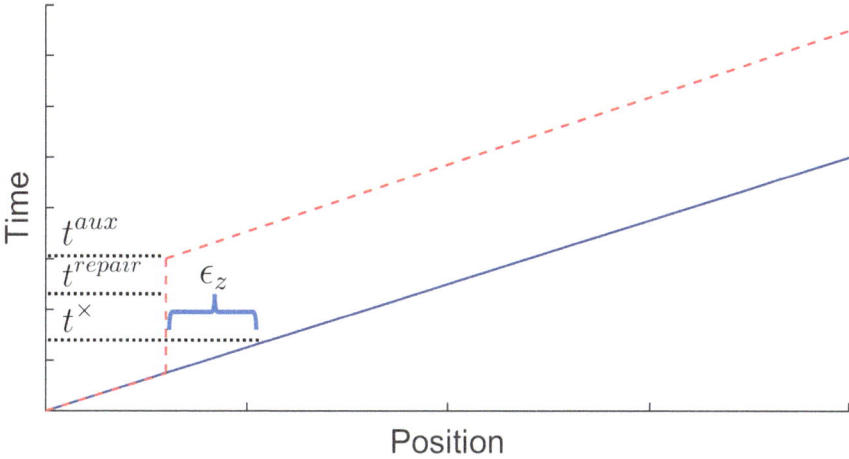

Fig. 3.5 Detection and rescheduling time for break down incidents

When engaging in rescheduling, two conflicting considerations come to the forefront. The first revolves around ensuring that the new trip plan achieves both optimal performance and maximum safety for all involved vessels. The second issue involves evaluating the number of vessels that will be impacted by the rescheduling process, including those that may need to adjust their trip plans or schedules in response to the incident. Minimizing the impact on the number of affected vessels is crucial, as significant disruptions or delays can adversely affect the overall efficiency and competitiveness of the port. Notably, substantial delays or operational disruptions can tarnish the port's reputation and discourage future business engagements [6]. Striking a balance between these conflicting concerns requires a comprehensive understanding of the specific incident and its potential effects on vessel operations. This often involves close collaboration with vessel operators and other stakeholders to gather and analyze data on vessel movements, traffic patterns, and other relevant factors.

Acknowledging the crucial equilibrium between optimizing performance and guaranteeing safety while protecting the reputation and competitiveness of the port, it becomes essential to investigate different rescheduling methodologies. In this research, we present three distinct rescheduling strategies: isolated rescheduling, global rescheduling, and multi-objective rescheduling. Subsequent subsections delve into each of these strategies, providing a detailed explanation and outlining their general formulations. These rescheduling methodologies are crafted to tackle the hurdles posed by arrival delays, speed reductions, and breakdown incidents in a methodical and efficient manner.

3.4.1   Isolated Rescheduling

In the case of the isolated rescheduling strategy, the primary goal is to devise a new optimal and feasible plan exclusively for the affected vessel, while keeping the trip plans of the remaining vessels (and consequently their arrival times at their respective destinations) unchanged. It is crucial to note that when addressing a breakdown incident, several specific considerations come into play during the rescheduling process. Firstly, it is assumed that, regardless of where the breakdown occurs, the affected vessel can be safely docked along the waterway's bank. This assumption relies on the channel's width being sufficient at any given point to accommodate the vessel without obstructing navigation or violating the crossing and head-on constraints outlined in the original plan. Secondly, it is essential to emphasize that the unaffected vessels within the fleet should continue sailing according to their original schedules until the new rescheduling process is implemented. This proactive approach minimizes the risk of additional delays or complications arising from schedule adjustments. Lastly, the rescheduling process must be carried out promptly to prevent further disruptions and ensure the ongoing efficient operation of all vessels. To adhere to operational constraints, vessels affected by such incidents should only resume navigation once indicated by a new decision variable denoted as t^{aux}.

The following optimization problem is formulated:

$$\min_{\mathbf{t}, t^{aux}, c} J_{sail}$$

$$\text{s.t.} \quad (4) - (12),$$

$$v_{i*} = \bar{v}_{i*} \text{ if speed reduction}, \tag{3.1}$$

$$t_{i,p} = t^{\dagger}_{i,p} \; \forall i \neq i^*, \forall p \in \mathbb{N}^Z_0, \tag{3.2}$$

$$t_{i*,p} = t^{\dagger}_{i*,p}, \; \forall p < p^{\times}_{i*}, \text{ if } i^* \in \mathcal{I}_u, \tag{3.3}$$

$$t_{i^*,p} = t_{i^*,p}^\dagger, \ \forall p \ge p_{i^*}^\times, \text{ if } i^* \in \mathcal{I}_d, \tag{3.4}$$

$$t_{i^*,p^\times+1} \ge t^{aux} + \Delta t_{i^*}^{aux} \text{ if } i^* \in \mathcal{I}_u, \tag{3.5}$$

$$t_{i^*,p^\times} \ge t^{aux} + \Delta t_{i^*}^{aux} \text{ if } i^* \in \mathcal{I}_d, \tag{3.6}$$

$$\text{in (10) and (11) substitute } t_{i^*,p^\times} \text{ by } t^{aux}, \tag{3.7}$$

$$t^{aux} = t^\times \text{ if delay or speed reduction}, \tag{3.8}$$

$$t^{aux} \ge t^{repair} \text{ if break down}, \tag{3.9}$$

where t^{aux} represents the new decision variable introduced to determine the optimal time for the affected vessels to resume sailing. In the case of speed reduction or arrival delay incidents, this time must coincide with the moment when the incident is detected and the optimal trip plan is rescheduled, as indicated in constraint (3.8). For break down incidents, this time should be greater than the moment when the affected vessel is repaired, as specified in constraint (3.9).

To prevent collision accidents, constraints (3.10) and (3.11) must be adjusted as specified in constraint (3.7). This issue is illustrated in Fig. 3.6, where a vessel sailing upstream (depicted in red) is affected by a breakdown incident. To avoid a collision with the downstream vessel (depicted in blue), t^{aux} must be greater than $t_{j,p}$. In the figure it can be seen how the affected vessel must wait until the other vessel leaves the section. Constraint (3.1) is introduced to modify the maximum speed at which vessel i can sail in the case of a speed reduction incident. Constraint (3.2) ensures that the new trip plans of vessels unaffected by the incident remain consistent with their original plans, where $t_{i,p}^\dagger$ represents the time when vessel i crosses boundary p according to the originally scheduled plan. Constraints (3.3) and (3.4) ensure that the crossing times of the boundaries already passed by vessel i^* are consistent with the original trip plans. Constraints (3.5) and (3.6) take into account the current

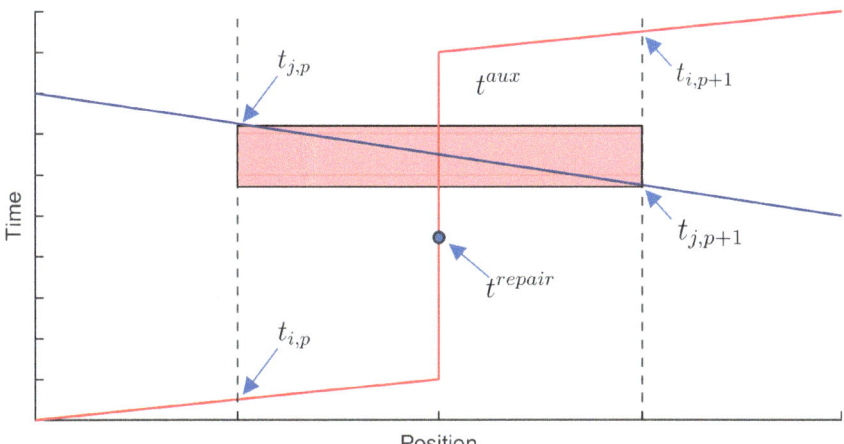

Fig. 3.6 Illustrative example of the problem associated with the crossing constraint following the repair of a vessel affected by a break down incident

position and the distance already covered by vessel i^*, where Δt_{i*}^{aux} represents the time required to sail from the vessel's location at the time of rescheduling (or the location of the waterway where vessels are positioned when they are allowed to resume sailing in the event of a breakdown or an unknown position incident) to the location of the next boundary while sailing at the maximum speed.

3.4.2 Global Rescheduling

The isolated rescheduling approach concentrates on adjusting the trip plan of the affected vessel while maintaining the plans of other vessels unaltered. Although this strategy aims to reduce disruptions to the original plans, there are instances where it could lead to an infeasible scenario. In simpler terms, it might not be feasible to devise a new trip plan for vessel i^* that complies with speed and encountering constraints without modifying the plans of other vessels.

To demonstrate this problem, take a look at Fig. 3.7, where we can see the paths of various ships navigating a theoretical waterway according to their planned trips. Imagine a situation where there are two ships, one moving upstream (depicted by the red line) and the other downstream (shown by the blue line). For these ships to safely pass each other without a collision, they must meet within the green-shaded area of the waterway. Now, consider a scenario where, at a particular point marked by a black star in both time and space, the upstream ship encounters a reduction in speed. The new path for this ship, now traveling at a lower maximum speed, is indicated by the red dotted line. Consequently, any potential encounter between these two ships can only happen in the region where the blue line intersects with the area shaded in red, which is not a practical solution. Hence, it becomes clear that finding a viable solution without altering the course of the downstream ship (blue line) is unattainable.

To reduce the risk of being unable to find feasible new trip plans, one solution is to modify the trip plans of all vessels when an incident occurs, a strategy referred to as

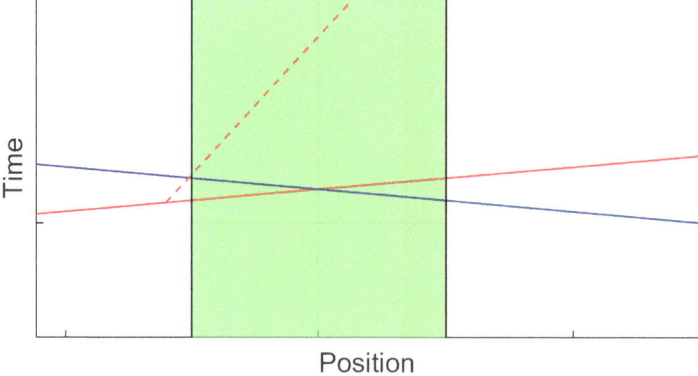

Fig. 3.7 Illustrative example of feasibility problems due to a speed reduction incident

global rescheduling. Similar to the preceding approach, this is achieved by solving the following optimization problem:

$$\min_{\mathbf{t}, t^{aux}, c} J_{sail}$$

$$\text{s.t.} \quad (4) - (12), \ (3.1), \ (3.5) - (3.9),$$

$$t_{i,p} = t_{i,p}^{\dagger}, \ \forall p < p_i^{\times}, \ \forall i \in \mathcal{I}_u, \tag{3.10}$$

$$t_{i,p} = t_{i,p}^{\dagger}, \ \forall p \geq p_i^{\times}, \ \forall i \in \mathcal{I}_d, \tag{3.11}$$

$$t_{i,p^{\times}+1} \geq t^{aux} + \Delta t_i^{aux} \ \text{if} \ i \in \mathcal{I}_u \backslash i^*, \tag{3.12}$$

$$t_{i,p^{\times}} \geq t^{aux} + \Delta t_i^{aux} \ \text{if} \ i \in \mathcal{I}_d \backslash i^*. \tag{3.13}$$

Constraints (3.10) and (3.11) ensure that the crossing times of the boundaries already crossed by each vessel i remain consistent with their original trip plans. Meanwhile, constraints (3.12) and (3.13) consider vessels i unaffected by the incident, accounting for the distance already traveled following the original plan at the time of the incident detection. Here, Δt_i^{aux} represents the time required for vessel i to sail from its location at the moment of rescheduling until it reaches the next boundary while maintaining maximum speed.

In contrast to the isolated approach, the global rescheduling strategy involves modifying the trip plans of a larger number of vessels. However, this broader scope may help alleviate some of the feasibility issues encountered in the isolated approach, as the optimization problem gains more flexibility to find suitable solutions.

3.4.3 Multi-objective Rescheduling

If the objective is to achieve an optimal solution that prioritizes both feasibility and minimal disruption to vessel schedules, a multi-objective approach can be utilized [7]. In this scenario, we consider that a vessel's trip plan is affected when its arrival time at the final destination deviates from the one specified in its original trip plan. To mitigate this discrepancy, a new penalty term is incorporated into the cost function of the global rescheduling problem. This penalty term is structured to penalize any disparities between the predicted arrival time based on the previous trip plan and the arrival time derived from solving the new rescheduling problem. This updated cost function is defined as follows:

$$J_{rescheduling} = \sum_{i \in \mathcal{I}_u} |t_{i,z} - t_{i,z}^{\dagger}| + \sum_{i \in \mathcal{I}_d} |t_{i,0} - t_{i,0}^{\dagger}|. \tag{3.14}$$

In this instance, it is crucial to highlight that only deviations in terms of arrival times for each vessel are taken into consideration. Alternatively, the entire trajectory could have been considered. The multi-objective rescheduling problem is formulated using a weighted sum approach, where the cost function is a linear combination of

J_{sail} and $J_{rescheduling}$. This method aims to achieve a balance between optimizing sailing efficiency to decrease operational costs and enhancing the system's adaptability and responsiveness to uphold safety and minimize disruptions. The multi-objective optimization problem can be expressed as follows:

$$\min_{\mathbf{t},t^{aux},c} J_{sail} + \eta \cdot J_{rescheduling} \qquad (3.15)$$

$$s.t. \quad (4) - (12), (14) - (26).$$

In this formulation, η serves as a single weighting parameter that influences different solutions depending on the prioritized cost. The obtained sailing and rescheduling costs for various η values represent Pareto optimal solutions for the multi-objective optimization problem, as outlined in [7]. These solutions form a Pareto front, constituting a collection of non-dominated solutions where no solution can be improved without compromising at least one of the objectives. Hence, a compromise must be reached between both objectives to determine the most appropriate solution. This selection process involves evaluating the trade-offs between the objectives and selecting the most favorable solution based on the decision-maker's preferences.

This method aligns with the constraints of the global rescheduling approach, meaning that the trip plans of all vessels can be adapted to achieve the desired objectives. Within the rescheduling cost, it is possible to integrate different weights to prioritize specific vessels. This flexibility empowers the decision-maker to assign greater importance to particular vessels, particularly those crucial for the smooth functioning of the transportation system. For example, vessels transporting essential goods or passengers might be assigned a higher weight in the rescheduling cost to ensure their punctual arrival at their destinations.

3.5   Case Study

In this section, we will carry out simulations to demonstrate the practical implementation of our proposed monitoring system and evaluate various rescheduling techniques applied to vessels navigating the Guadalquivir River. The same configuration of the river of that employed in the previous section is considered. To evaluate the proposed monitoring and rescheduling approach, a series of experiments were conducted using randomly generated scenarios. Each scenario is defined by the arrival times t_i^{ready} of vessels ready to enter the waterway. The sequence of arrivals at each end of the waterway is generated using a Poisson distribution with a characteristic parameter of λ. Both sequences commence at the initial time t^{init}. Vessel beam sizes are categorized as small, medium, or large, quantified as 1, 2, and 4, respectively, following the information provided by the Port of Seville. Each and every parameter used to characterize the different simulation scenarios has been chosen to demonstrate the main properties of the proposed approach. Throughout all experiments conducted in this section, a uniform distribution was employed to determine the beam size for each vessel. To assess the proposed rescheduling strategies, experiments were designed to introduce arrival delays, speed reductions, and breakdown incidents for the vessels.

For arrival delay incidents, each vessel may experience such an incident with a probability of p_{delay}. In such cases, a random arrival delay, following a uniform distribution between 0.5 and 1 h, is applied. For speed reduction incidents, each vessel may encounter a speed reduction incident with a probability of p_{speed}. In these instances, the maximum speed at which each vessel can sail is multiplied by a factor v_i, which is randomly generated from a uniform distribution ranging between 0.5 and 0.9. The timing of the incident is randomly selected during the first hour after the vessel enters the waterway, using a uniform distribution. Finally, for breakdown incidents, vessels may be randomly affected by breakdown incidents with a probability of ρ_{break}. The timing of the incident occurrence is randomly selected from a uniform distribution within the first three hours of simulation. The duration required to repair a vessel affected by the incident is randomly drawn from a uniform distribution between 1 and 2 h.

To demonstrate the need for rescheduling, we first show how the likelihood of encountering arrival delays, speed reductions, or breakdown incidents impacts the frequency of accidents in the waterway. An accident is defined as when two vessels meet in a section of the waterway where the width is inadequate to accommodate both vessels, posing a significant risk of collision.

A set of 100 scenarios was generated with parameters $N = 8$ and $\lambda = 0.5$ h, covering arrival delays, speed reductions, and breakdown incidents. These scenarios varied in their probabilities, denoted as p_{delay}, p_{speed}, and p_{break}, spanning from 0.1 to 0.9. For each scenario, optimal trip plans were devised for every vessel, followed by a simulation where vessels traversed each boundary at their scheduled times specified in the plans. In case of arrival delay incidents, late arriving vessels adhered to the speed prescribed by their original plans. For speed reduction incidents, affected vessels (referred to as A.V) sailed at the lower speed between their original plan and the new maximum speed post-incident. If a breakdown occurred, vessels resumed their voyage at the speed dictated by the optimal plan after repairs. The average number of accidents was recorded as a function of varying probabilities p_{delay}, p_{speed}, and p_{break}, and the results are depicted in Figs. 3.8, 3.9, and 3.10, respectively. These figures illustrate the direct relationship between incident probabilities and an increase in collisions.

Our experimental results unveil a noticeable connection between the probabilities of arrival delays, speed reductions, and breakdown incidents, and the average frequency of accidents. With higher probabilities of these incidents, the likelihood of accidents also escalates. Furthermore, our findings highlight that even minor probabilities of incidents can lead to notable accidents, underscoring the vital necessity of continuous monitoring and rescheduling to ensure vessel safety and prevent accidents.

As indicated before, incidents are identified by leveraging information from the AIS, which involves real-time comparisons between each vessel's current location, denoted as $z_i^{\dagger}$, and its expected location based on its optimal trip plan, denoted as $z_i(t)$. If the disparity exceeds a specific threshold, ϵ_z, which in this instance has been set at $\epsilon_z = 5$ km/10, the incident is flagged as detected. In Fig. 3.11, the anticipated

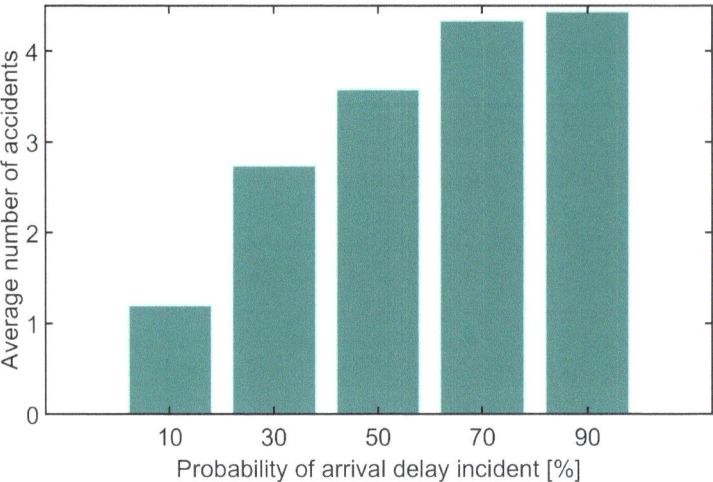

Fig. 3.8 Average number of accidents as a function of the probability of arrival delay incident

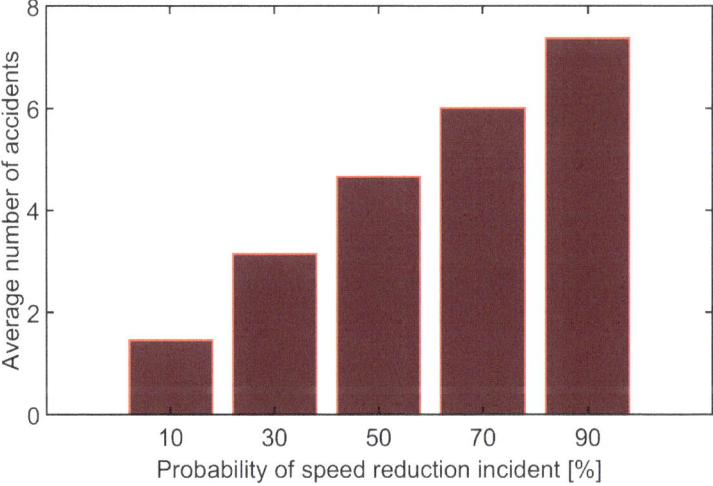

Fig. 3.9 Average number of accidents as a function of the probability of speed reduction incident

trajectory and speed of a vessel according to its optimal trip plan (depicted by the blue lines) and the actual trajectory and speed can be observed, taking into account a speed reduction incident at time 120 that reduces its maximum speed by a factor of $v_i = 0.5$ (depicted by the red lines). Notably, the incident occurs at time 120, yet it is not until time 127 that the incident is detected, resulting in a detection time lag of 7 h/10.

Following that, a series of simulations were executed to demonstrate the effectiveness of the monitoring system in identifying incidents. Each simulation involved

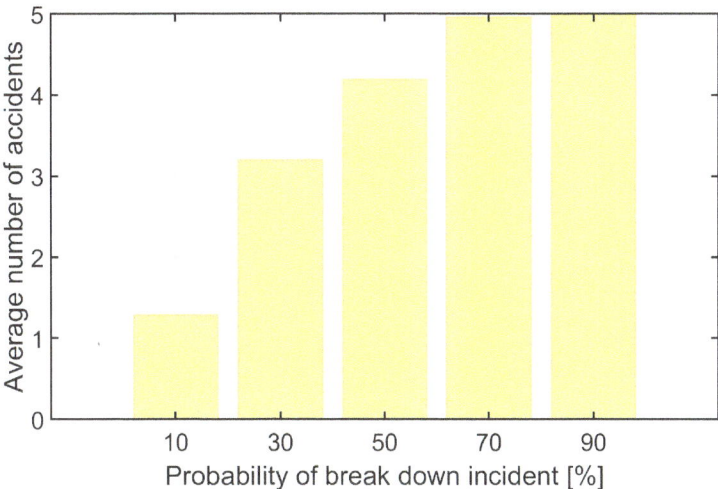

Fig. 3.10 Average number of accidents as a function of the probability of break down incident

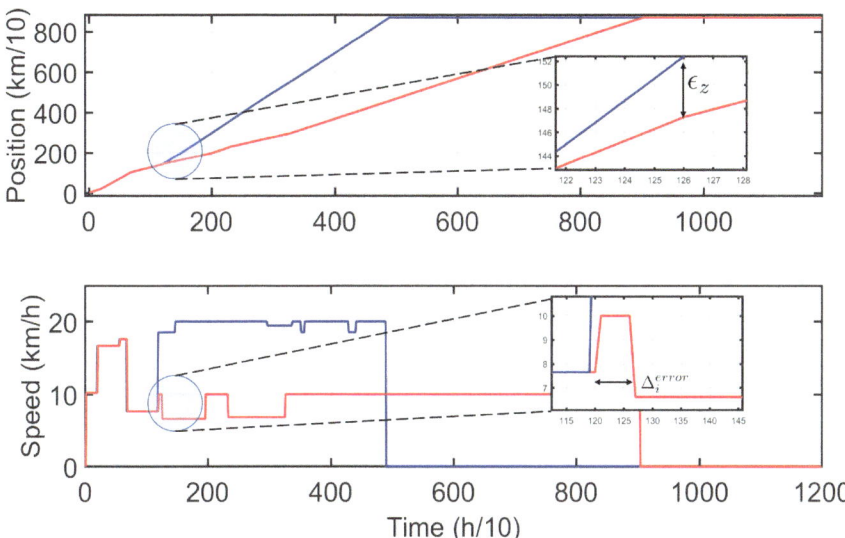

Fig. 3.11 Incident detection

the arrival of 8 vessels at an average rate of one vessel every 2 h ($\lambda = 0.5$ h). In the initial simulation, we induced arrival delays by randomly postponing the arrival of the first vessel for a duration ranging from 30 to 60 min. In the subsequent simulation, speed reduction incidents were emulated by randomly selecting a reduction factor between 0.5 and 0.9 for the first vessel during the first hour of navigation. In both scenarios, a monitoring zone of 5 km/10 was presumed ($\epsilon_z = 5$ km/10).

For the third simulation, breakdown incidents were introduced by randomly determining the time of the first vessel's arrival in the waterway. It was assumed that the vessel would undergo repairs within a time span of 2–3 h, following a uniform distribution.

The results from our simulations are depicted in Figs. 3.12, 3.13, 3.14, 3.15, 3.16 and 3.17, illustrating the outcomes of arrival delay, speed reduction, and breakdown incidents, employing both isolated and global rescheduling approaches. In

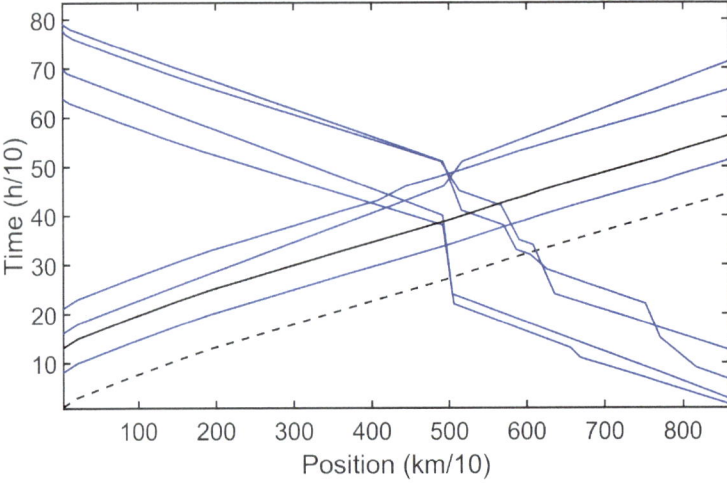

Fig. 3.12 Example of isolated rescheduling for an arrival delay incident

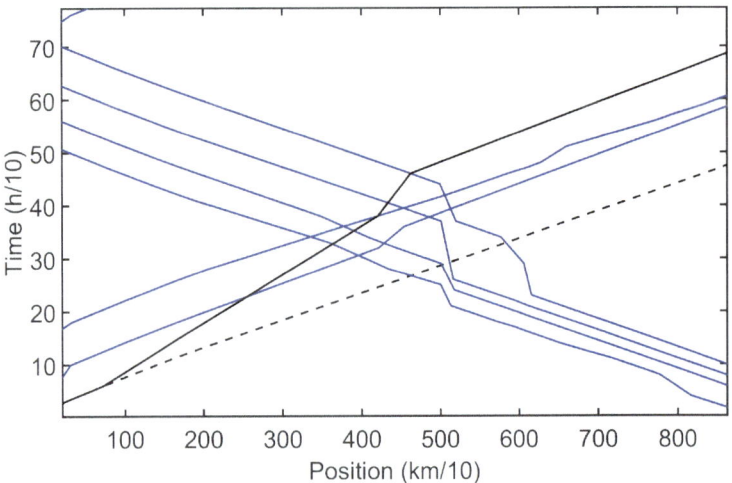

Fig. 3.13 Example of isolated rescheduling for a speed reduction incident

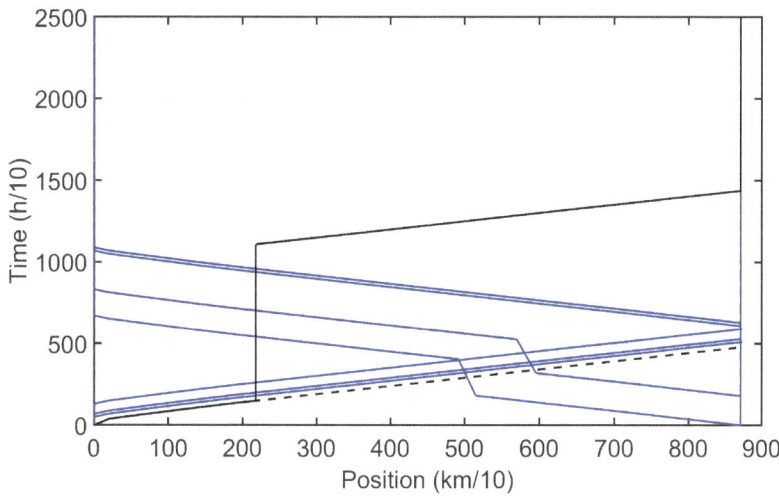

Fig. 3.14 Example of isolated rescheduling for a break down incident

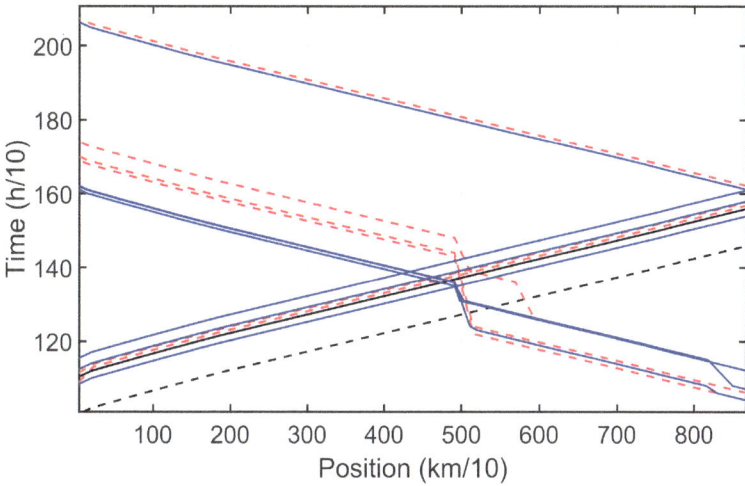

Fig. 3.15 Example of global rescheduling for an arrival delay incident

these visuals, the expected trajectories are denoted by dotted red lines, while the actual trajectories post-rescheduling are shown by continuous blue lines. For the vessel directly impacted, the expected trajectory is represented by dotted black lines, and the continuous black lines portray the actual trajectories after rescheduling. Our findings suggest that when employing the isolated rescheduling strategy, the trajectories of all vessels, except the affected one, remained unchanged. This is attributed to the strategy's focus on rescheduling exclusively for the impacted vessel, leaving the others to adhere to their original routes and schedules. Conversely, with the

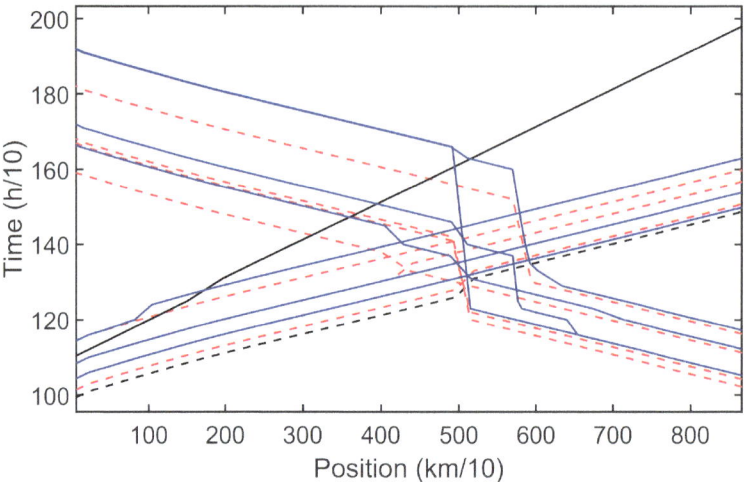

Fig. 3.16 Example of global rescheduling for a speed reduction incident

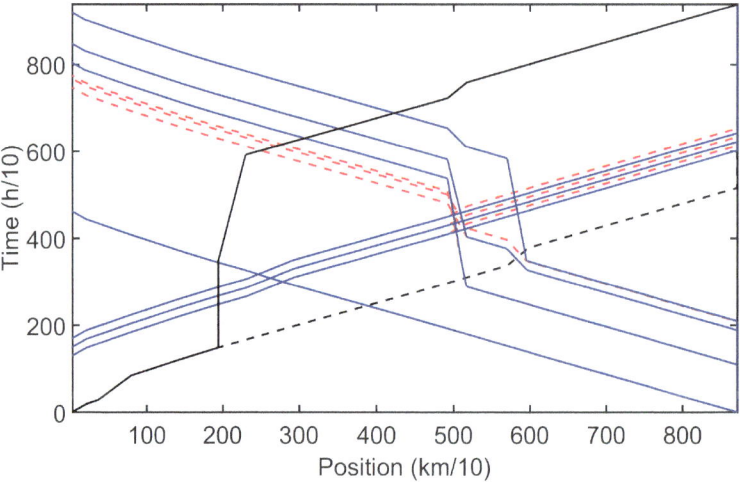

Fig. 3.17 Example of global rescheduling for a break down incident

application of the global rescheduling strategy, the trajectories of most vessels were affected post-rescheduling.

To elucidate the balance between sailing expenses and rescheduling expenses, we formulated a series of scenarios, each involving $N = 8$, $\lambda = 1$, and $\epsilon_z = 5$ km/10 for every incident type. In this study's findings, we utilized average values derived from the first 100 feasible scenarios. Within each scenario, multiple simulations were executed, wherein different rescheduling strategies were employed: isolated rescheduling, global rescheduling, and multi-objective (M.O) rescheduling, with varied values of η spanning from 0.01 to 100. The results obtained from each simulation, if attainable, encompass the number of affected vessels and the absolute deviation between the arrival time at the final destination as anticipated by the original optimal plan and the arrival time at the final destination post-rescheduling. We categorize a vessel as impacted by rescheduling if this absolute deviation surpasses 2 h/10.

In cases of arrival delay incidents, the initial vessel entering the waterway faces a randomly generated delay spanning from 1 to 3 h. The mean outcomes gleaned from our experiments are consolidated in Table 3.1. The findings demonstrate clear constraints on both the average number of vessels impacted by rescheduling and the average deviation between arrival times. Specifically, the former is minimized, while the latter is bounded above by the number of affected vessels when employing isolated and global rescheduling strategies, respectively.

In the case of speed reduction incidents, a comparable procedure is adopted. Here, the initial vessel entering the waterway encounters a speed reduction incident within the first hour of navigation, with its maximum speed reduced by a factor of $v_i = 0.6$. The mean findings are showcased in Table 3.2. Notably, both the average number of vessels impacted by rescheduling and the average deviation between arrival times are limited. Specifically, the former is constrained from below, while the latter is capped from above by the number of affected vessels when utilizing isolated and global rescheduling strategies.

Table 3.1 Simulation results for arrival delay incidents

Rescheduling strategy	Arrival delay incidents		
	J_{sail} [h/10]	$J_{resscheduling}$ [h/10]	A.V
Global	41.12	5.72	6.51
M.O ($\eta = 0.1$)	41.44	5.69	6.28
M.O ($\eta = 0.5$)	42.28	3.12	3.56
M.O ($\eta = 1$)	44.19	0.79	2.89
M.O ($\eta = 10$)	44.36	0.61	2.23
M.O ($\eta = 100$)	44.75	0.61	1.21
Isolated	45.34	0.58	1

Table 3.2 Simulation results for speed reduction incidents

Rescheduling strategy	Speed reduction incidents		
	J_{sail} [h/10]	$J_{rescheduling}$ [h/10]	A.V
Global	43.09	2.98	5.12
M.O ($\eta = 0.01$)	43.80	3.07	4.39
M.O ($\eta = 0.1$)	44.05	2.67	3.87
M.O ($\eta = 1$)	44.86	2.25	2.56
M.O ($\eta = 10$)	45.03	2.11	1.34
M.O ($\eta = 100$)	45.49	2.04	1.28
Isolated	46.76	2.03	1

Table 3.3 Simulation results for break down incidents

Rescheduling strategy	Break down incidents		
	J_{sail} [h/10]	$J_{rescheduling}$ [h/10]	A.V
Global	31.97	20.42	6.76
M.O ($\eta = 0.1$)	32.55	21.29	6.30
M.O ($\eta = 0.5$)	38.24	15.17	5.41
M.O ($\eta = 1$)	47.81	6.65	4.93
M.O ($\eta = 10$)	47.99	6.34	3.71
M.O ($\eta = 100$)	46.23	5.63	2.69
Isolated	46.89	4.65	1

When breakdown incidents occur, a vessel is randomly selected to encounter a breakdown. It is anticipated that this vessel will experience the incident within the initial three hours of the simulation, with the breakdown duration ranging from 3 to 4 h. The average outcomes from this simulation are outlined in Table 3.3, illustrating once more how the results yielded by the multi-objective rescheduling method are bounded above and below by those attained through the global and isolated rescheduling strategies, respectively.

Regarding arrival reduction incidents, the optimization problem remained consistently feasible for both isolated and global rescheduling strategies. However, for speed reduction incidents, the optimization problem was feasible in 16.5% of instances with isolated rescheduling and in 66.3% with global rescheduling. Conversely, in the case of breakdown incidents, the optimization problem was always feasible for both isolated and global rescheduling strategies.

3.6 Conclusions

This chapter introduced a real-time monitoring system and a multi-objective rescheduling strategy designed to rearrange vessel schedules in natural inland waterways when incidents involving any of the vessels in the original plan occur. Experiments have been conducted, and comparisons made with three alternative rescheduling strategies: isolated rescheduling, global rescheduling, and multi-objective rescheduling. The results underscore the effectiveness of the multi-objective rescheduling strategy, demonstrating superior performance in terms of fewer vessels being affected by rescheduling and less delay in arrival times. This strategy allows for a flexible trade-off between the number of affected vessels and the delay incurred by the rescheduling strategy. The value of η plays a crucial role, directly impacting the number of vessels influenced by rescheduling and the discrepancy in arrival times between the original and rescheduled trip plans. As η increases, the number of affected vessels decreases, and the difference in arrival times diminishes. This highlights the importance of carefully selecting the η value to strike the right balance between minimizing disruptions and mitigating delays. In essence, the multi-objective rescheduling strategy offers a promising approach to enhance the safety and efficiency of vessel navigation in natural waterways, particularly in the face of unexpected incidents.

References

1. R.T. Marler, J.S. Arora, The weighted sum method for multi-objective optimization: new insights. Struct. Multidiscip. Opt. **41**(6), 853–862 (2010)
2. J. Nadales, D.M. de la Penã, D. Limon, T. Alamo, Real-time monitoring and optimal vessel rescheduling in natural inland waterways. IFAC-PapersOnLine **56**(2), 7880–7885 (2023)
3. S. Kim, H. Kim, Y. Park, Early detection of vessel delays using combined historical and real-time information. J. Oper. Res. Soc. **68**(2), 182–191 (2017)
4. Ø.Ø. Dalheim, S. Steen, Uncertainty in the real-time estimation of ship speed through water. Ocean Eng. **235**, 109423 (2021)
5. A. Galierikova, T. Kalina, J. Sosedova, Threats and risks during transportation of lng on European inland waterways. Transp. Probl. **12**(1), 73–81 (2017)
6. M. Acosta, D. Coronado, M. Mar Cerban, Port competitiveness in container traffic from an internal point of view: the experience of the Port of Algeciras Bay. Marit. Policy & Manag. **34**(5), 501–520 (2007)
7. G.P. Rangaiah, *Multi-objective Optimization: Techniques and Applications in Chemical Engineering*, vol. 5 (World Scientific, 2016)

Chapter 4
Practical Implementation of Scheduling Tools

Abstract In this chapter, we present a practical implementation of a software tool designed for scheduling vessels in natural inland waterways, following the requirements imposed by the port authorities of the Guadalquivir River, an inland port located in the south of Spain. The implemented system focuses on optimizing the planning of cargo vessel journeys in inland waterways. This, along with the integrated library of functions, equips navigation planners with a comprehensive solution for optimizing vessel routes while considering the constraints imposed by varying depths and encounter situations. By considering real-time data and environmental conditions, it enables the calculation of safe crossing windows (tubes) for vessels to pass through critical points along their journeys, thereby minimizing the risk of accidents and ensuring safe navigation. This implementation represents a significant step forward in the field of inland navigation planning, offering a valuable resource for improving logistics operations in these challenging environments. The codes with the different functionalities for the implementation of the safe tube finding algorithm as well as the optimization problem for calculating the optimal trip plans of all vessels have been distributed through an open-source repository.

4.1 Introduction

Efficient navigation planning in natural waterways presents a distinctive set of challenges, necessitating adaptive solutions that account for the dynamic interplay of changing environmental conditions and operational constraints. Addressing these complexities is crucial, particularly in regions with specific requirements, such as the Guadalquivir River, an inland port situated in the southern reaches of Spain. Throughout the chapters of this book, the development of algorithms for the management of uncertainty induced by the tidal effect and the planning of navigation in natural inland waterways has been carried out. This concluding chapter delves into the implementation of a software tool meticulously crafted to meet the exacting demands outlined by the port authorities of the Guadalquivir River. It represents the culmination of the work developed in this book, aiming to provide a practical implementation of the algorithms developed in this book and tailored to the current needs of the Port of Seville.

Our focus revolves around the optimization of cargo vessel journeys within inland waterways, a task demanding a nuanced approach that integrates seamlessly with the intricacies of real-time data and fluctuating environmental conditions. To this end, we implement the scheduling methodology proposed in Chap. 2. The implemented system, along with its integrated library of functions, stands as a testament to innovation in navigation planning. Specifically tailored to the unique constraints of the Guadalquivir River, this software tool empowers navigation planners with a comprehensive solution.

At its core, the system concentrates on refining the planning of cargo vessel routes, where considerations extend beyond the standard navigational challenges to encompass factors such as varying depths and encounter situations. By embracing a forward-thinking approach, our software enables the calculation of safe crossing windows as proposed in Chap. 2, allowing vessels to navigate critical points along their journeys with minimized risk and heightened safety standards. The main contribution of this chapter lies in the practical implementation of a safe tube finding method and the scheduling algorithm for finding the optimal trip plans for a set of vessels in natural inland waterways, effectively minimizing the risk of accidents and ensuring secure navigation.

To foster collaboration and further advancements in the field, we have taken a commendable step by distributing the codes encompassing diverse functionalities through an open-source repository. This not only encourages transparency but also invites the collective expertise of the community, thereby propelling the evolution of navigation planning tools and strategies.

4.2 Tool Overview

As presented in Chap. 2, the optimal navigation plan must ensure that each vessel reaches its destination within the specified time frame while also guaranteeing compliance with all operational constraints, including those imposed by the dynamic effect of tides. The idea is that the tool serves not as an autonomous guidance system but as a navigation aid tool. In accordance with current navigation regulations in most channels of this type, it is assumed that vessel control and maneuverability are the responsibility of the pilot, an expert navigator for the channel. The pilot is responsible for adjusting the vessel's speed to adhere to the plan. This means that, for a given optimal plan t_{ip}, there is not a unique trajectory, but rather a set of possible paths to follow that adhere to the plan. The choice of following one trajectory over another will always depend on the navigation criteria adopted by the pilot, which will define the profile followed by the vessel.

The overall framework of the proposed tool for the implementation of the desired functionalities is shown in Fig. 4.1. This tool is always managed by the waterway manager. The objective, rather than finding the individualized optimal plan for each of the vessels, is to find the global optimum, always taking into account that the established contracts between port authorities and the various vessels regarding entry

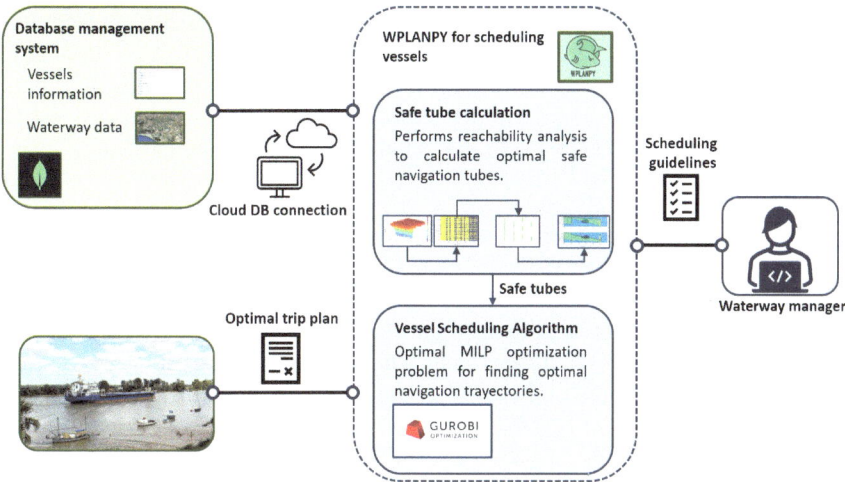

Fig. 4.1 General architecture of the proposed solution for scheduling vessels in inland waterways

and stay times in the channel must be respected. Non-compliance with these could not only affect the port's reputation but also lead to contractual penalties.

In this way, the planning tool receives from the waterway manager the priority of each of the vessels that have requested to navigate through the channel and are waiting in one of the parking areas at both ends of the channel for permission to access it, as well as the instructions for carrying out optimal navigation, always under the control of the expert pilot in river navigation. Taking into account the scheduling guidelines provided, the developed planning tool is responsible for calculating optimal plans for each of the vessels collectively. This is done in a two-layer structure. In the first layer, the tool calculates the set of safe tubes for each of the vessels based on the estuary characteristics, vessel data, and tidal predictions. These safe navigation tubes are defined as a collection of temporal windows, each associated with the various waypoints along the river, indicating the time intervals during which it is safe to cross each of them, ensuring that the depth restriction imposed by the river's maximum depth at these locations, taking into account the bathymetric profile and tide predictions, is not violated. To this end, the following parameters are needed:

- The depth map, which takes into account both the bathymetric profile of the river and the tide prediction for the considered time window. This map defines the time instances at which the tidal restriction is violated at each position along the waterway.
- The draft of each vessel.
- The maximum speed at which each vessel can navigate.
- The maximum speed allowed at each section of the waterway.
- The distance of each section.
- The estimated time of arrival to the corresponding extreme of the waterway.

The output of this layer is the set of safe tubes for each vessel, indicating the crossing windows when each vessel can cross each waypoint safely. In the second layer, using the set of safe tubes for each vessel, the tool plans the voyages of each vessel by solving a global optimization problem that simultaneously considers all the vessels. To accomplish this task, the following parameters are needed:

- Depth map.
- Maximum speed of each vessel.
- The beam of each vessel.
- Width of each section.
- The minimum and maximum times when vessels can cross each section boundary according to the set of safe tubes calculated in the previous stage.
- Optimality criteria to be defined by the performance index of the optimization problem.

The output of this layer is the set of optimal plans for all vessels, i.e., the time instants when each vessel must cross each waypoint in the waterway. All the information required by the tool to carry out the planning is stored in a cloud-based database, which can be accessed in real time through a query from the tool. Once the optimal navigation plans have been found, they are provided to the different vessels.

4.2.1 Tool Schema

As explained in the previous section, the first step in the planning process, given the information pertaining to each of the vessels and the state of the river, is to identify the set of possible safe navigation corridors that each of the vessels can follow. It is important to note here that the calculation of the corridors depends on the draft of each of the vessels, as well as the number of waypoints considered when calculating the safe crossing windows. Generally, while the corridors ensure safe passage through each of the windows, there is no guarantee about what happens in the middle of the section, and it is the responsibility of the pilot to navigate the vessel to avoid depth-related issues. The greater the number of sections into which the river is divided, the higher the safety of the identified corridors, but this comes at the expense of increased computational cost, which grows exponentially with the number of sections. Additionally, a pruning procedure can be adopted to ensure safe navigation in the middle of the different sections. For an in-depth study of how these corridors can be found using the reachability algorithm, the reader is referred to Chap. 2, where a detailed treatment of this topic is discussed in detail.

The first step is to identify the set of time windows when it is safe to cross each waypoint in the waterway. Once this is done, the set of safe corridors for each of the vessels is determined through a recursive reachability analysis between the windows in each section, taking into account both the maximum allowable speed in the channel and the maximum speed that each of the vessels can achieve. Navigating through the river following any of these safe tubes ensures the existence of at least one trajectory within them that complies with the depth restriction in every location, as well as the maximum allowable speed in each section of the waterway.

Once the tool has found the set of safe corridors for each vessel, the next challenge is to solve an optimization problem. This takes the form of a MILP and determines the optimal plan for each vessel, taking into account a specific economic criterion. The economic criterion to be optimized can vary from finding the set of safe crossing times that minimize transit waiting times, to identifying those times that ensure the channel is traversed within the maximum time period while adhering to established limits, thereby reducing the average navigation speed. This reduction in speed is generally associated with lower fuel consumption and, consequently, a reduced environmental impact of the transport process. It will be the decision of the waterway manager to determine which economic criterion to adopt when conducting the planning. Furthermore, the planning process may prioritize certain vessels over others by assigning a higher weight to the costs associated with them in the optimization problem.

In the following section, we focus on detailing each of the modules used by the tool, their relations, and the technologies employed to implement them.

4.3 Software Architecture

In this section, we provide a detailed explanation of each of the modules that make up the proposed planning tool. Each and every one of them has been implemented using Python programming language. All the information required for the planning is stored in a MongoDB document-oriented NoSQL database. The optimization problem has been implemented using the MIP modeling library [1]. For the resolution of the optimization problem, the Gurobi solver has been chosen, which requires prior installation on the computer or service where the tool is executed. The choice of solver to be used by the tool is open, and any solver capable of solving mixed-integer problems can be used [2]. The dependency diagram of each and every module of the tool is represented in Fig. 4.2.

4.3.1 Cloud Database

Instead of using a traditional table-based database to store the different data required to calculate the set of safe tubes and the optimal plan for each vessel, a document-based NoSQL database has been utilized. Document-based NoSQL databases like MongoDB provide a schema-less data model, allowing for dynamic and unstructured data storage. This flexibility makes them well-suited for applications with evolving data requirements and complex data structures [3]. In this type of database, all information is stored in documents, which are essentially sets of data stored in a key-value format. Moreover, each of the documents stored in the database has a unique identifier, simplifying the process of retrieving information by knowing the document's identifier and the key of the data you want to access. The proposed

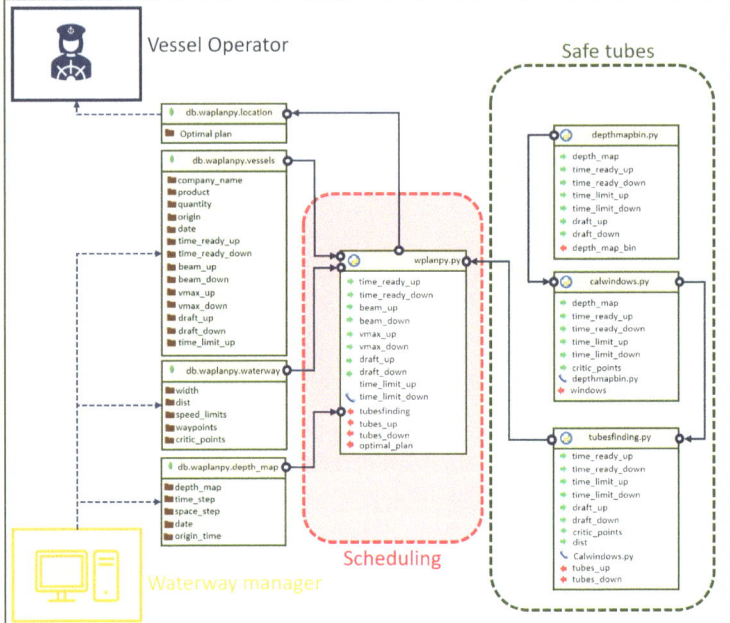

Fig. 4.2 Dependency diagram of back-end codes

tool uses three document schema to store information, one for storing the specific information of each of the vessels, one storing the data of the waterways, where the information corresponding to the characteristic of the waterway is stored in a single arraying containing the information for all sections, and finally, a document for storing the information of the depth map to be employed by the safe tube calculation algorithm. In addition, the same database is employed to store the optimal plan result of the optimization problem (see Fig. 4.2). These documents and their contents are shown in the Listings 4.1 to 4.4.

Listing 4.1 JSON document for storing data of each vessel

```
{
"id": <vessel reference>,
"time_arrival": <time of arrival>,
"date": <date of arrival>,
"origin": <origin port>,
"company": <company name>,
"product": <product name>,
"quantity": <quantity of product>,
"beam": <beam>,
"draft": <draft>,
"max_speed": <maximum speed>
}
```

Listing 4.2 JSON document for storing waterway information

```
{
"waypoints": <waypoints>,
"dist": <distances>,
"width": <widths>,
"speed_limits": <speed limits>
}
```

Listing 4.3 JSON document for storing waterway information

```
{
"depth_map": <depth map>,
"time_step": <time step>,
"space_step": <space step>,
"date": <date>,
"origin_time": <origin time>
}
```

Listing 4.4 JSON document for storing waterway information

```
{
"optimal_plan": <optimal plan>
}
```

The first document is employed to store the information of each vessel. A different document is needed for each of the vessels. This document contains the identifier (id) of each vessel, which is automatically generated by the database, time (time) and date (date) when the vessel is supposed to arrive at the waterway, the origin port (string), which determines whether the vessel navigates upstream or downstream, the name of the operating company (string), the quantity in tons (float) of the transported product, and the operational parameters of each vessel, namely, their widths (integer), drafts (float), and maximum speed (float).

The second document is used to store the information of the waterway. In this case, a single document is employed to store the information of all sections in the waterway. This document contains the length (float), width (integer), and speed limits (float) of each section, and set of waypoints (array of integer) and critical points (array of integers).

The third document is used for storing the information about the depth map. This document contains the depth map (array of float), the time and space steps (float), which indicate the time and space discretization employed to build the depth map, the date (date), and the initial time (time).

The last document is employed to store the resulting optimal trip plans (array of float). This document can be accessed by the vessel's pilot at any time by accessing the cloud-based database.

4.3.2 Safe Tubes Calculation

As it was previously described, the set of safe tubes for each vessel must be calculated before the optimization problem for finding the optimal plan is performed.
We denote these sets $\mathcal{K}_i$, for each vessel $i \in \mathcal{I}$. When function *tubesfinding.py* is
called from *wplanpy.py*, which receives as input the information about the waterway,
the information about the different vessels, and the depth map, the first step is to
find the set of safe crossing windows for each of the sections, taking into account
the bathymetric profile, tide prediction, and the draft of each of the vessels. The
procedure for this is as follows. First, the *calculatewindows.py* function is called,
which receives as input the tide prediction map and vessel information. Within this
function, the tide map is transformed into a binary map, where each of the elements
in the matrix describing the map is assigned a value of 0 or 1 depending on whether
the draft of the vessel is less than or greater than the depth indicated by the timestamp
and position given by the depth map. To this end, the function *depthmapbin.py* is
called. Using the binary depth map, the crossing windows at each of the waypoints
are found by means of a processing procedure that finds the borders of the regions
in the depth map where the depth constraint is violated. The set of safe crossing
windows is denoted as Q_p, for each waypoint $p \in \mathbf{N}_0^Z$.

   In principle, we could calculate the set of safe tubes as causal combinations of these
windows. Nonetheless, the quantity of potential safe crossing tubes would increase
exponentially as the number of sections increases. Furthermore, while any of these
permutations would ensure that depth limitations are upheld at the different waterway
boundaries, there is no guarantee regarding potential occurrences while traversing
each section, considering depth and speed restrictions. To solve this problem, the
proposed tool integrates the reachability analysis-based tree algorithm proposed in
Chap. 2 to search for feasible connections between safe crossing windows, ensuring
the existence of at least one trajectory that allows transitioning from one waypoint to
another without violating speed or depth constraints. This is achieved by locating the
minimum and maximum points within the input crossing window, such that a line with
a constant slope equal to the inverse of the maximum allowable speed in the section
connects with the output crossing window without violating the depth restriction at
any of the intermediate points in the section. An example of safe tubes for a vessel
sailing upstream is shown in figure, where the yellow lines represent the safe crossing
windows at each waypoint of the waterway and the dotted blue line represents the time
limit. An example of safe tubes calculated using the tool is represented in Fig. 4.3.

4.3.3 Planning Module

The main module of the tool is named *wplanpy.py*. This module takes as input data
read from the cloud-based database, which includes information about each of the
vessels. It then returns, as output data stored back in the database, the optimal trip
plan for each vessel.

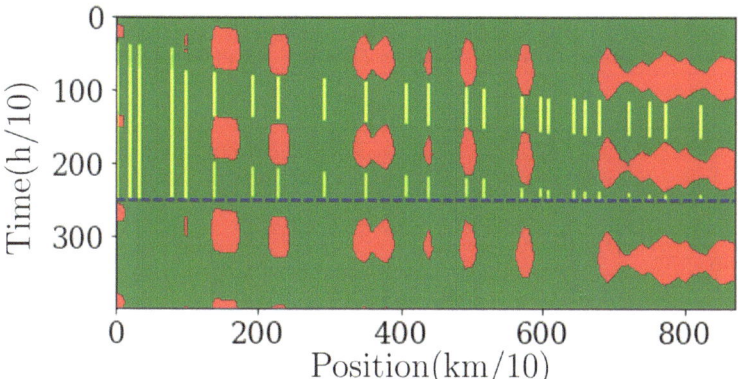

Fig. 4.3 Example of safe crossing tubes for a 7.5 m draft vessel. the safe crossing windows are represented in yellow. The dotted blue line represents the time limit

The behavior of this module is as follows. First, once the input data have been read from the cloud database, the tool calculates the set of safe tubes for each vessel. To do this, the function *tubesfinding.py* is called. This module receives as input data the information about the different vessels, the information describing the waterway, and the depth map, and it returns, as output data, the set of tubes to be used by the optimization problem. Using the set of safe tubes, and the information of each vessel, the module build the optimization problem proposed in Chap. 2 using MIP Python library.

Once the optimal plans for all vessels have been calculated, the tool stores them in the database so that each of the vessels can read them directly. Additionally, a graphical representation of the path that each of the vessels should follow is provided, assuming that the vessels travel at a constant speed to reach each of the different waypoints at the time indicated by the optimal plan. Note that this trajectory may vary depending on the navigation criteria applied by the pilot of each vessel.

4.4 Simulation Results

In this section, we test the behavior of the developed tool through a case study applied to the Guadalquivir River. To do this, we create a scenario where, for confidentiality reasons, the values of each of the vessels as well as the bathymetry and tidal data have been randomly altered. The information relative to the depth map is an array of float stored in the document *db.wplanpy.depth_map* taking the form of a 1398 (time steps) by 871 (space steps) array with a sampling time $\Delta t = 6$ min and sampling space $\Delta d = 100$ m. This information is provided by a series of depth sensors located along the river.

We consider a scenario in which a total of 8 vessels intend to access the channel during the same operating window, half of them sailing upstream and the other half sailing downstream. All of them have their origin at the Port of Seville (S) for

Table 4.1 Vessel parameters

Origin	Beam	Max. speed (km/h)	Draft (m)	Time arrival
Vessels downstream				
S	2	22	7.5	6:00
S	1	20	6.5	6:12
S	2	22	7	6:24
S	2	18	7.5	6:54
Vessels upstream				
C	2	20	7.8	6:03
C	1	20	6	6:10
C	1	18	6	6:20
C	1	17	6.5	6:30

the downstream-bound vessels and at the stuary of the Guadalquivir River in the town of Chipiona (C) for the upstream-bound vessels. The values that characterize each of the vessels are detailed in Table 4.1, which show the information extracted from the document *db.wplanpy.waterway*. Note that the beam value, instead of its actual measurement, receives a discrete value in the range 1–4, indicating the size as small, medium, large, or very large, following the common practice in this type of waterways. The values used for the description of the waterway are the same that those detailed in Chap. 2, which are extracted from the documents stored in the collection *db.waplanpy.vessels*. The optimal plan obtained using the proposed tool for this scenario are shown in Fig. 4.4. This optimal trip plan is stored in the document

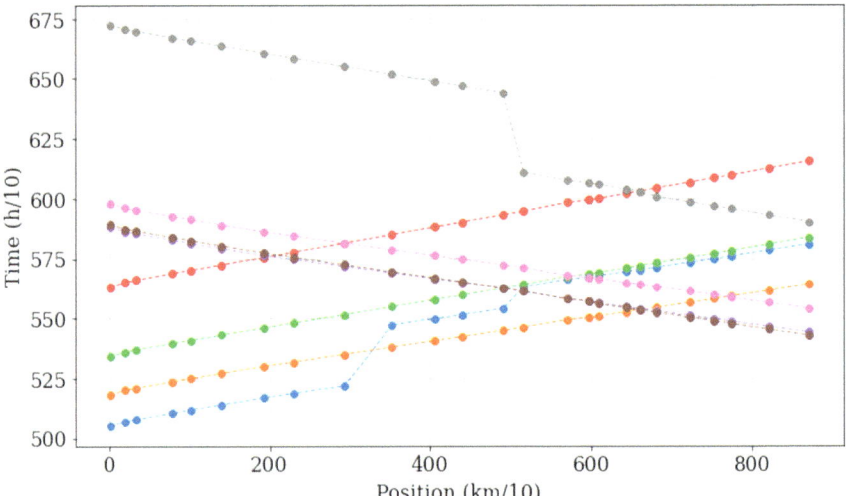

Fig. 4.4 Simulation results

db.wplanpy.location. For the calculation of the optimal plan, the same priority weight has been given to each of the vessels. Instead, different weights could have been used to prioritize the transit and waiting of certain vessels.

All the functionalities detailed in the chapter for the calculation of the safe navigation tubes and the resolution of the optimization problem for obtaining the optimal plans are available in the following open-source repository under MIT license: https://github.com/JMNadales/WPLANPY.

4.5 Conclusions

In this chapter, we introduced an innovative tool for computing optimal navigation plans in natural inland waterways. Beyond addressing various operational constraints, our approach uniquely considers the dynamic impact of tides in the planning process to mitigate the risk of accidents. This tool represents a significant advancement in navigation planning methodologies for complex natural waterway environments, offering a more comprehensive and safety-oriented approach. The tool operates in two primary phases. Firstly, leveraging using the depth map for a specific time frame, it computes a set of safe navigation tubes. These tubes delineate temporal windows during which it is secure to traverse each section of the waterway, ensuring the existence of at least one viable trajectory for safe canal navigation. Secondly, building upon the established set of safe tubes and considering additional operational constraints, the tool calculates a set of secure navigation plans tailored to each individual vessel. The tool has been validated through practical implementation in a real-world case study centered on the Guadalquivir River, a natural inland waterway located in the south of Spain.

References

1. H.G. Santos, T. Toffolo, Mixed integer linear programming with python. Accessed April 2020
2. R. Anand, D. Aggarwal, V. Kumar, A comparative analysis of optimization solvers. J. Stat. Manag. Syst. **20**(4), 623–635 (2017)
3. B. Jose, S. Abraham, Performance analysis of NoSQL and relational databases with MongoDB and MySQL. Mat. Today: Proc. **24**, 2036–2043 (2020)

Chapter 5
Conclusions

The culmination of the individual studies within this book has yielded comprehensive insights and innovative solutions aimed at enhancing the planning and navigation of vessels in natural inland waterways, with a particular focus on the challenging environment of the Guadalquivir River. The application of the proposed methodologies, as demonstrated in each of the chapters of this book, would result not only in an improvement in the safety of navigation but also in an increase in the productivity and efficiency of the process in economic and social terms.

In Chap. 2, a two-step optimization-based approach was proposed for scheduling and planning vessels, considering time-varying depths in the waterway is presented. Extensive simulations in the Guadalquivir River demonstrated the applicability of the formulation in solving planning problems for varying scenarios. The proposed solution not only improved upon existing first-arrived first-served strategies but also showcased computational tractability, paving the way for its potential application as a management tool for the Port of Seville.

In Chap. 3, we introduced a real-time monitoring system and a multi-objective rescheduling strategy, proving its effectiveness through experiments and comparisons with alternative strategies. The multi-objective strategy demonstrated superior performance in terms of minimizing the number of affected vessels and reducing transit and waiting times, offering a flexible trade-off that can be fine-tuned using a weighting parameter. This multi-objective rescheduling strategy stands as a promising approach to enhance safety and efficiency in vessel navigation, particularly when faced with unexpected incidents.

Finally, in Chap. 4, we presented a software tool for computing optimal navigation plans in natural inland waterways, addressing operational constraints and dynamically considering tides to mitigate the risk of accident. The tool, operating in two phases, calculates safe navigation tubes and secure navigation plans for each individual vessel. The practical implementation on the Guadalquivir River validated the tool's functionalities, and all code related to this work is available in an open-source repository.

J. Moreno Nadales et al., *Optimal Vessel Planning in Natural Inland Waterways*,
SpringerBriefs in Applied Sciences and Technology,
https://doi.org/10.1007/978-3-031-64744-4_5

Collectively, the chapters in this book represent a contribution to the field of inland navigation planning, offering not only individual advancements but also the potential for a holistic and integrated approach to improve safety and efficiency in complex natural waterway environments.

The findings and insights derived from this research work provide a foundational understanding of scheduling and rescheduling in natural inland waterways. However, as the scope of problems in natural inland waterways continually presents challenges and opportunities, this work presents potential directions for further exploration and refinement, aiming to contribute to the ongoing evolution of knowledge in the field of scheduling and navigation within natural inland waterways. Some of these possible research lines for future improvement are listed below:

- Throughout this book, different algorithms have been developed for the scheduling and rescheduling of vessels in natural inland waterways. Moving forward, a primary avenue of future work lies in the implementation of these tools on real scenarios. The translation of these algorithms from theoretical constructs to operational solutions represents a critical step in bridging the gap between academic research and real-world maritime challenges. This transition toward practical implementation aligns with the overarching goal of enhancing the efficiency and effectiveness of vessel scheduling and navigation in natural inland waterways. As the industry demands solutions that seamlessly integrate with existing frameworks, the forthcoming focus on implementation underscores the commitment to making tangible contributions to the field and addressing the pressing needs of maritime stakeholders.
- In the field of business process models, a noteworthy future direction involves expanding existing models, mainly focused on discrete events, with dynamic elements. The envisioned research proposes the integration of dynamic scheduling models into business process simulations, specifically targeting port management in natural inland waterways. By incorporating dynamic factors like fluctuating water levels and real-time traffic, this extension aims at enhancing the fidelity of the simulations, offering a more realistic and accurate representation of the challenges faced in natural inland waterways and port operations. Such an approach has the potential to contribute significantly to optimizing decision-making processes and improving overall operational efficiency in the intricate domain of inland waterway logistics.
- The effective integration of natural inland waterways channel planning with other critical operations, such as the Berth Allocation Problem and Quay Crane Scheduling Problem, is of paramount importance for optimizing the overall efficiency of port management. These operations are closely interconnected, with channel planning playing a central role in facilitating smooth vessel access to berths and terminals. Seamless coordination between channel planning and onshore tasks, such as berth allocation and crane scheduling, ensures proper synchronization, minimizing wait times, and optimizing resource utilization. This strategic integration not only enhances operational productivity but also contributes to cost reduction and maximizes port capacity, crucial elements for the overall success of logistics operations in inland waterway environments.